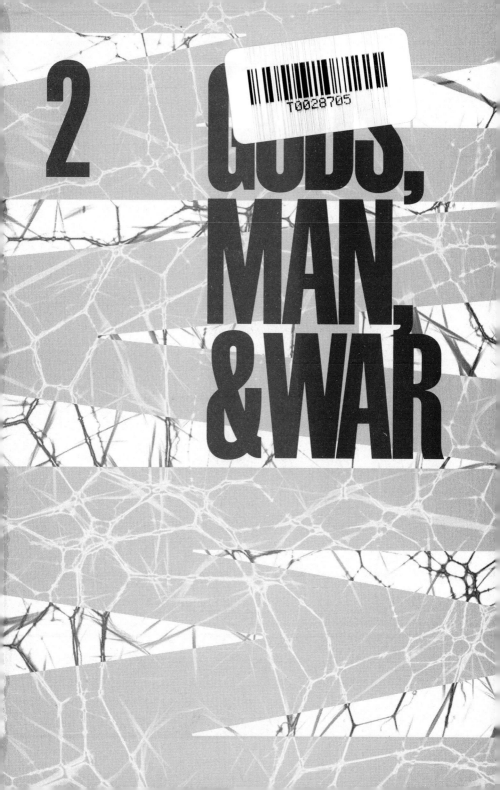

2

GODS, MAN, &WAR

MAN

VOLUME 2

GODS, MAN, & WAR

An official
Sekret Machines
investigation
of the UFO
Phenomenon

By
Tom DeLonge
with
Peter Levenda

To The Stars, Inc.
1051 S. Coast Hwy 101 Suite B, Encinitas, CA 92024
ToTheStars.Media
To The Stars… and *Sekret Machines* are trademarks of *To The Stars, Inc.*

Managing Editor: Kari DeLonge
Copy Editor: Kate Petrella
Consulting: David Wilk
Book Design: Lamp Post

Manufactured in the United States of America

ISBN 978-1-943272-37-2 (Hardcover)
ISBN 978-1-943272-42-6 (Trade Paperback Edition)
ISBN 978-1-943272-38-9 (eBook)

Distributed worldwide by Simon & Schuster

To Julian and Holden, Star Travelers

PETER'S ACKNOWLEDGMENTS:

I would like to acknowledge the team from To The Stars Academy of Arts & Science: some very busy individuals who generously gave of their time and attention to look over the manuscript and suggest changes or additions. This list must include Jim Semivan, Hal Puthoff, Garry Nolan, and Chris Mellon. I would also like to thank Christopher "Kit" Green and Eric Davis for their kind comments and support. And, of course, this project would never have gotten off the ground at all had it not been for Tom DeLonge's vision and drive. From *Sekret Machines* to The Stars, it has been—and remains—a wild ride!

Thanks also to Kari DeLonge for keeping us all in check and on schedule and for her many helpful suggestions as well.

I would also like to thank Whitley Strieber for his friendship and guidance on all things Visitor. I have had many long conversations with Whitley about the Phenomenon, and they always give me much to think about and provide a valuable perspective on something that defies easy explanation.

Also many thanks to friends and family: people I have had to avoid in order to do the research and the writing (and the thinking) on this most difficult subject. Writers are antisocial beings anyway, but Tom had me pondering the philosophy and psychology of machines in a manner I had never considered before and set off a trail of thought that has been proceeding unbroken and uninterrupted for more than three years now, making me seem like an absent-minded troll most of the time. My apologies!

And it should go without saying: any errors of fact or interpretation here are solely my own.

Peter Levenda

CONTENTS

Prologue

PROLOGUE

If anyone says they have the answers, they're fooling themselves.

We don't know the answers but we have plenty of evidence to support asking the questions. *This is about science and national security.* If America doesn't take the lead in answering these questions, others will.

12:59 P.M. - 16 Dec. 2017

–Senator Harry Reid
(emphasis added)

I N A TWEET PUBLISHED ON DECEMBER 16, 2017—THE day the news broke nationwide that Tom DeLonge's To The Stars Academy of Arts & Science (TTSAAS) had served as the vehicle for the release of two official Department of Defense videos showing the existence of UFOs—former Nevada Senator Harry Reid admitted there was a secret program at DOD to investigate UFOs, and that "This is about science and national security."

In the first book of this trilogy—*Sekret Machines: Gods*—we explored the ancient cultural evidence for Contact. This ancient evidence took the form of religious and spiritual texts concerned with beings from the heavens and the relationship between these beings and the impetus all across the globe to create civilizations around architectural devices that were believed connected to the "gods" and to the heavens. These were also associated with divine-human contact and the possibility of immortality and altered states of consciousness. In ancient Sumer and Egypt—among other cultures—this meant a kind of spiritual *technology* (at least, that is how it would be understood today).

In this book we will pick up where we left off, but by taking a page from Harry Reid and his UFO project. We will look at the implications of the UFO where science is concerned. In our third and last book—*Sekret Machines: War*—we will examine the national security implications.

While we feel that it is problematic to separate the spiritual, scientific, and security fields from each other when discussing the UFO Phenomenon, we also realize that we as human beings have been trained to think this way, at least in the past two hundred years or so since Newton. We have compartmentalized knowledge to such an extent that the chance of a real Renaissance Man or Woman rising from among us has been drastically diminished. While we can't hope to rectify that situation, we do intend to show how a multidisciplinary approach to the problem of the UFO Phenomenon will yield the best results and will advance the cause of global knowledge and security, if not a Renaissance of the human spirit.

Further, you will notice that we will spend a great deal of time in this volume referring to the contactee and abductee aspect of the Phenomenon. To many in the UFO/UAP (unidentified aerial phenomena) community, this aspect is the least credible or unworthy of being "lumped in" with the purely physical component of the

Phenomenon. However, as recent research by Dr. Garry Nolan and his colleagues suggests, there are physical effects in contactees that can be observed and traced on a biological level; and as Dr. John Mack observed in 1994, abductees are suffering from a form of PTSD, indicating that they had experienced some kind of psychological trauma. This combination of physiological and psychological evidence makes it incumbent upon us to integrate the contactee/abductee experience with the UFO/UAP experience to some extent, and we have attempted to do so here.

One of the characteristics of UFOs that is common in the popular literature is that these "sekret machines" represent a technology that is far in advance of our own. There is a flaw buried deep within that argument, however. Our technology is advancing by leaps and bounds, and has been since World War II. Our achievements in genetics, physics, computer technology, and artificial intelligence have now reached what once were science-fiction levels of sophistication and complexity, and there is no sign that this trend will slow down in the foreseeable future. So the question has to be reformulated to take these advances into consideration. If the UFO Phenomenon represents a technology more advanced than our own, *advanced by how much?*

In ancient times, the technology was seen as magical and otherworldly. There was no *Journal of the American Medical Association* in those days; no *Scientific American*. The scientific journals of the time are interpreted today as religious or magical or spiritual texts. The ancient peoples were trying to come to terms with phenomena that seemed incredible, even impossible for humans to achieve. Today that same technology is seen as simply a few levels higher than our own. We could not fly in ancient times. We could not communicate over long distances. We could not see the trans-Saturnian planets. Today, however, we are gradually reaching the same level of technological expertise as the purported "ancient aliens." What does that say about the "alien" culture that already has that technology?

More important, what does that say about *us*?

▼   ▼   ▼

To The Stars Academy identified several research areas for their focus in the coming years. We have selected a few of them for investigation here, in order to explain the reasons they were chosen and their implications for the study of the Phenomenon. These are consciousness, the brain-computer interface, and telepathy, along with genetics.

Genetics is chosen because of proposals by Francis Crick and others that the RNA and/or DNA molecules were designed and seeded onto this planet deliberately. We find that there is anecdotal evidence to support this idea in several divinatory systems from Africa, China, and elsewhere, as well as in ancient "religious" systems of Egypt and India, among others. We will examine the theories put forward by scientists concerning the genetic code and amplify them with reference to these systems.

Consciousness studies is the hot new field combining neurobiology, genetics, and quantum mechanics. There are several important theories of consciousness that deserve our study, but most important are their implications for understanding both the UFO/UAP Phenomenon as well as human-"alien" contact.

One of the following sections is concerned with the brain-computer interface and thus with artificial intelligence, robotics, and cyborgs. We are close to developing machines that will seem as sentient as human beings but with none of the physical or emotional vulnerabilities. When that happens, we will be forced to ask ourselves some far-reaching questions, such as what it means to be human. We will be forced to ask whether those who pilot or control the UFOs are nothing more than robots or cyborgs, the kind we are in the process of building ourselves. That might have sounded fanciful only thirty years ago, but today there is a very real possibility that human-created androids of some type will be piloting our spacecraft outside the solar system to more distant star systems. Doesn't that automatically make us consider that the "aliens" reported by our experiencers and abductees are nothing other than the same technology we ourselves are developing?

We will also look at telepathy, taking into consideration the scientific research done in the area of remote viewing and also bringing in what we have already learned about consciousness and the brain-computer interface. As many abductees report being in telepathic contact with the "aliens" who abducted them or performed some kind of experimentation on them, is it possible for us to employ the techniques of mental communication in reverse? Is there a component on the microlevel, the non-Newtonian level, that can be exploited to allow us to communicate in new ways with other human beings or even with those nonhuman actors connected with the UFO Phenomenon?

These and many other possibilities will be explored in the pages that follow. The information and data we employ will come exclusively from experts in the scientific and engineering communities, from their published work and from personal interviews. The arguments proposed will come largely from ourselves and our advisers, some of whom you already know.

The conclusions will be exclusively your own.

Tom DeLonge
Peter Levenda

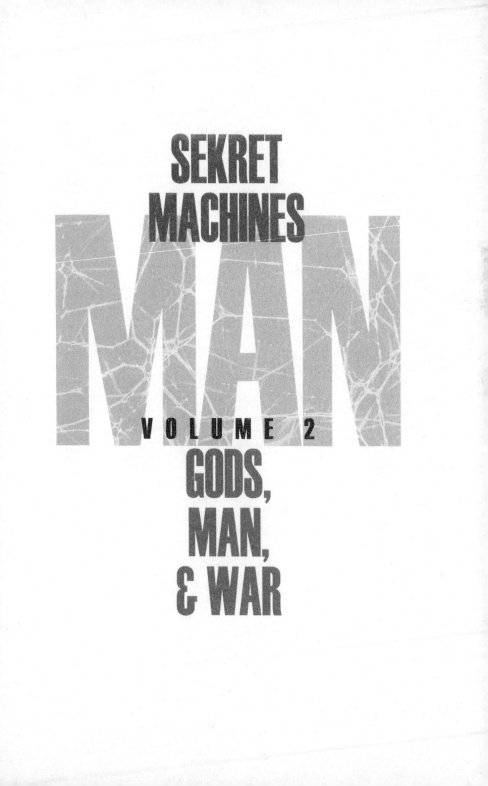

SEKRET MACHINES

MAN

VOLUME 2

GODS, MAN, & WAR

SECTION ONE

▼

GENETICS AND THE EXTRATERRESTRIAL HYPOTHESIS

INTRODUCTION TO SECTION ONE

Either the machine has a meaning to life that we have not yet been able to interpret in a rational manner, or it is itself a manifestation of life and therefore mysterious.

> – Garet Garrett, *Ouroboros: Or the Mechanical Extension of Mankind*

We are survival machines–robot vehicles blindly programmed to preserve the selfish molecules known as genes. This is a truth which still fills me with astonishment.

> – Richard Dawkins, *The Selfish Gene*

AT THIS TIME THERE HAS BEEN RENEWED PUBLIC INTERest in robots, artificial intelligence, and the idea that consciousness may not be restricted to biological structures but may be a characteristic of advanced forms of machines.

We are moving gradually to the point at which we will ask these questions seriously, with regard to a potential and perhaps threatening future where we co-exist with machines we have created but that have turned *on* us, or turned *away from* us.

If we—*homo sapiens sapiens*—are on the verge of developing sentient machines, does that not imply that perhaps what we have experienced in terms of the Phenomenon may be just that: sentient machines, and not "extraterrestrial biological entities" (EBEs)?

All of those harrowing, nightmarish experiences claimed by alien abductees—the kidnapping from their bedrooms in the dark of night, the experimentations, the physical intrusion of tools, equipment, and the like—may be due to machines inspecting what they believe to be other machines: us. After all, we feel no remorse, no twinge of conscience, when we tinker with the innards of a personal computer, a smartphone, or a television set, or work on a car or a boat or a plane. They are all machines, inorganic devices we have designed and built and mass-manufactured. But as we approach ever more closely to designing, building, and mass-manufacturing better and faster and smarter machines, we are reaching a point at which it will become virtually impossible to distinguish a machine from a human being.

Are the Visitors—the ubiquitous aliens of our fantasies, our experiences, and our radar traces—actually machines? Have they simply reached that point ahead of us, by fifty or a hundred years or so? Are we in danger of not being able to recognize that which is alive and that which is merely a simulacrum of life? Or is the danger more visceral than that: have we become increasingly unsure of what the dividing line is between life and nonlife?

Indeed, have the Visitors?

▼　　▼　　▼

There are now robots that look amazingly human, with synthetic skin that feels real (even warm) to the touch, and with artificial intelligence (AI) capabilities such that they can converse with us, even sing. They do not have the lurching, staggered moves of Hollywood robots but are designed to operate more smoothly. While it is impossible to mistake one of these robots for an actual human today, in a few years the distinction will be hard to make.

If we were to land on an alien planet and confront a being there, would we be absolutely certain that what we saw was not an organic life form but an artificial creation, left there by some other civilization or perhaps created on that planet itself, years or even centuries earlier?

Ask yourself one question in particular: with all the evidence that has accrued over the past seventy years concerning close encounters, alien abductions, and the like, why have no alien children ever been spotted? Why do the Visitors we have seen not age or show signs of aging? Could it be that they are not organic life forms that are born, age, and die, but are machines that are created in the form we see them?

If so, were they created by organic beings like ourselves, or by other machines?

These sound like fantastic ideas—or they would have, even fifty years ago—but these are questions we now have to consider, carefully and dispassionately. They lead us to question the nature of life and of consciousness. These are the very questions that are raised by the Phenomenon, and our inability to answer them thus far has made it virtually impossible to identify, characterize, and explain the Phenomenon. In order to understand *them* we have to do a better job of understanding *ourselves*.

According to the Biblical account, human beings were artificial creatures made by God for no particular reason we can discern. Thus we are already a kind of machine, made "in the image and likeness" of God, just as we are creating robots in our own "image and likeness."

As the scientist-philosopher Loren Eiseley wrote:

> Man reached his mind into the emptiness to seize upon the machine; with it willingly or not, he seized hold of the future. Ever since, he has been trying to free himself from the embrace of the future.[1]

Is it that bad? Or have we learned to "stop worrying and love the Bomb" as the tagline for the film *Dr. Strangelove* would have it? Or conversely, as the tagline for the *X-Files* movie asks us, do we "Fight the Future"?

▼   ▼   ▼

Many contactees and abductees report that experiences with the Visitors are terrifying, not least because there seems to be no empathy coming from them toward their human subjects, whom they treat as things to be taken apart and studied. For some reason, we are naïve enough— politically, anthropologically, psychologically—to expect a certain degree of "humanity" from them, which is nowhere assured or indicated. In fact, their affect seems uniformly detached; one might say "machine-like." Is this what we ourselves are creating? Is this the end point, the natural conclusion, of the much publicized "Convergence"?

> Perhaps *Homo sapiens*, the wise, is himself only a mechanism in a parasitic cycle, an instrument for the transference, ultimately, of a more invulnerable and heartless version of himself.[2]

THE SUCCUBUS IN THE SAUCER

I would never say, yes, there are aliens taking people. I would say there is a compelling powerful phenomenon here that I can't account for in any other way, that's mysterious.

– Dr. John E. Mack, Professor of Psychiatry,
Harvard Medical School[3]

There exists a global phenomenon the scope and extent of which is not generally recognized. It is a phenomenon so strange and foreign to our daily terrestrial mode of thought that it is frequently met by ridicule and derision by persons and organizations unacquainted with the facts.

– Dr. J. Allen Hynek, Professor of Astrophysics,
Harvard Observatory[4]

O
NE ASPECT OF THE PHENOMENON THAT DESERVES A closer look is the spectacle of alien abduction and the type of close encounters that result in human beings being adversely affected by contact. This is probably the most prickly of all the themes we encounter in a study of the literature and through our personal interactions with experiencers. It is easily the most controversial aspect of the Phenomenon, the one with virtually no corroborating eyewitness evidence and which leaves few discernible traces except in the psyches of the individuals involved, as well as (at times) some bizarre physical symptoms. The stories they tell us are outlandish and bizarre. It is far easier to accept UFO sightings and even speculate about crashed

saucers than it is to believe that someone was taken from their bed in the middle of the night, beamed aboard a spaceship, physically abused (which would be the legal definition of being prodded, poked, probed, and even inseminated) and then returned to that same bed as if nothing had happened. And, in fact, UFO skeptics insist on exactly that: nothing happened.

A Case in Point

While stories of alien contact have appeared regularly since the early 1950s, when George Adamski claimed an ongoing relationship with a Venusian, the more frightening alien abduction scenario came to public attention with the case of Barney and Betty Hill, who claimed they were abducted from their automobile while on a lonely road at night, driving back to their home in New Hampshire from a trip to Montreal. It was the evening of September 19–20, 1961, and while it would become the most famous case of alien abduction, it went unreported in the press until 1964.

The facts of the case are not as clear as either the believers or the skeptics would insist. Sometime around midnight the Hills saw what they believed to be a UFO along Route 3 in northern New Hampshire (south of the Old Man of the Mountain site). They stopped the car to get out and take a closer look. Barney Hill, initially irritated by his wife's insistence that they stop and look at what was probably only an aircraft, became frightened by what he saw through a pair of binoculars he kept in the car. The object was large, had windows, and what appeared to be people of some kind staring out the windows at him. Barney began to feel a distinct sense of terror, and ran back to the car.

The couple was about to get back on the highway and head for home when they heard a strange beeping sound they could not identify. At this point they seemed to have lost consciousness.

For two hours.

They "came to" after hearing a second set of "beeps." Barney was still driving, and they were a long way from their home near Portsmouth. They reported that they did not feel as if they were fully conscious until they saw a sign saying "Concord 17 miles."

They had no idea what had happened. They felt odd, out of sorts. Once home they began to do things that didn't make a lot of sense, such as Betty putting her luggage near the back door of the house for no particular reason. Their watches had stopped and wouldn't start again. Barney, for reasons he could not explain, felt he had to go into the bathroom and examine his genitals, but there was nothing unusual about them.

They had had an experience that was at once terrifying and strange.

People are terrified all the time for perfectly acceptable reasons: war, crime, disease, natural disasters, and all "the slings and arrows of outrageous fortune." But then there is a category of terror that is not so easily defined or characterized, for it seems to come out of nowhere. It manifests during times of what Dr. J. Allen Hynek—during a speech about UFOs before the United Nations on November 27, 1978—referred to as "high strangeness."

The accounts given by abductees/contactees are similar in several details, but a common denominator is their sense of complete "otherness" just before, during, or after a contact. It is a complex emotion that has elements of fear, dislocation, and awe, and is often accompanied by physical symptoms such as a feverish feeling, nausea, dizziness, trembling, difficulty breathing, etc. Psychiatrists have attempted to characterize these feelings as stemming from night terrors or from various levels of sleep deprivation or suggestibility contributing to a kind of waking nightmare or hallucination. This is a somewhat disingenuous position to take. Of course there are various psychological states that show similarities to the alien abduction experience, such as sleep paralysis or hypnagogic hallucinations that can occur in the moments just before falling asleep, but to assume that those coming forward with experiences of alien abduction are all suffering from sleep paralysis is as far-fetched as any other explanation and is evidence of bias on the part of the observer rather than deception on the part of the contactee.

Another objection to the reality of the alien abduction experience is the assumption that the details of the experience are all culturally conditioned. In other words, if one has seen a movie about aliens, certain details from that film help to create the delusion that one is experiencing actual alien abduction. While it is entirely possible that perceptions of the experience may be filtered through a cultural lens, that does not negate the initial experience itself; it only negates the way in which it is described by the experiencer. It's like the road show of a popular musical: the actors may be different and the costumes not identical to the original, but the songs are all the same, as is the overall plot.

> A cardinal mistake, and a source of great confusion, has been the almost universal substitution of an interpretation of the UFO Phenomenon for the phenomenon itself.[5]

This is the situation that currently bedevils the UFO community: the preponderance of interpretations, speculations, and UFO "gurus" whose contributions often far outstrip the evidence. As we have seen in *Sekret Machines: Gods*, human beings perceive this Phenomenon in different ways depending on their cultural conditioning, but the Phenomenon still exists as an event separate from normal waking (and sleeping) reality and is perceived accordingly. The seemingly outlandish tales that are told by experiencers are the sincere efforts of human beings who have no working vocabulary for this type of encounter, much like the mystical literature one encounters in many cultures in which words fail the mystic as he or she struggles to communicate the ineffable. Because our scientific and technological trajectory has left the field of consciousness (and its related disciplines, such as religion, mysticism, dreams, the psyche, etc.) behind, we increasingly rely upon scientific and technological language to describe our world. When we apply that terminology to the Phenomenon, we are left sounding vaguely foolish. Or just vague. That is more the problem of language—which *is* culturally conditioned—than it is of the reliability of the experience itself.

We can determine the outlines of this sense of "high strangeness" in the actions of Barney and Betty Hill in the days and weeks following their experience on the New Hampshire highway. Betty Hill, believing they had seen a UFO, went to the library the day after the incident and borrowed a book on the subject by Donald E. Keyhoe, a major in the Air Force and an important figure in the field of Ufology. Eventually she contacted Major Keyhoe and told him as many details as she could recall, including her husband's statement that he had seen humanoid passengers aboard the UFO. She also phoned Pease Air Force Base on September 21, 1961, to report her sighting, leaving out those details she felt would make her sound insane (such as the humanoid occupants that Barney saw). She received a return phone call the following day from Major Paul Henderson, who filed his report on September 26, believing that the couple had mistaken the planet Jupiter for a UFO. That report eventually was included in the Project Blue Book files of the Air Force.

On October 21, 1961, the Hills were visited by a member of the National Investigations Committee on Aerial Phenomena (NICAP), which was the leading civilian UFO investigative body at the time and was headed by Major Keyhoe. (NICAP was cofounded by Major Keyhoe and Thomas Townsend Brown in 1956 and lasted until about

1980.) Barney admitted that there were elements of the story that eluded him, and that he thought he was suppressing some memories. Of the two, Betty Hill seemed more at ease with the experience, while Barney was more anxious and troubled.

Then Betty started having unsettling dreams involving a spaceship, medical exams conducted by humanoid figures aboard the ship, and other elements that are now familiar to many concerning alien abduction. The dreams were vivid and took place over five consecutive nights during the first week of October 1961.

It would not be until March 1963 that the Hills mentioned this experience in a public forum, which happened to be their local Unitarian Church. This aspect is important, for it shows that the Hills were not intent on creating a publicity furor over this event. They had remained quiet about it for eighteen months. In fact, even after this minor disclosure, they remained reticent and only agreed to hypnosis sessions to retrieve repressed memories in January 1964.

Under hypnosis, the Hills—who were hypnotized separately—began recalling many more details about the contact than they had consciously remembered. Their therapist, Dr. Benjamin Simon, was not a believer in alien abduction, although he had witnessed UFOs before. Regardless, he initially thought that Barney Hill's anxiety was caused not by alien abduction but by some form of childhood trauma. In the end, his conclusion was that the alien abduction scenario was the manifestation of a new, previously unknown form of psychological disorder, and he left it at that.

However, it seems that at least one tape of the Hills' hypnosis sessions was leaked to the press. In October 1965—and thus a full four years after the alleged abduction—the story made the front page of the Boston newspapers. Suddenly, the Hills were an internationally famous couple.

The Barney and Betty Hill story attracted a great deal of attention, some of it sympathetic, but obviously that was not the intention of the Hills, who sought no publicity at all. By the mid-1960s, however, and with the publication of John G. Fuller's book about the event,[6] the Hills had become poster children for Ufology in general and alien abduction in particular. Betty Hill enjoyed the celebrity, while Barney remained troubled by it. Some psychiatrists offered the opinion that the proximate cause of his anxiety was sensitivity over his relationship with Betty. This was the early 1960s, in New Hampshire, and Barney Hill was a black man married to Betty, a white woman. However, the racial theory was not a conclusion reached by their therapist, Dr. Simon.

Debunkers entered the fray and tried to suggest that the details came not from their own subconscious minds or from an actual encounter but from memories of early television shows—like *The Twilight Zone* or *The Outer Limits*—or from pulp magazines or articles about UFOs; in other words, cultural contamination, a phenomenon that seems to exist only in connection to UFOs and alien abduction scenarios. There is no evidence to support this contention, of course, but it does make debunkers feel

Hopkins was a nationally recognized and lauded painter who was awarded everything from a Guggenheim fellowship to a grant from the National Endowment for the Arts, eventually winning a full membership at the National Academy of Design (1994). He wrote and lectured on art for decades, but it was his interest in the Phenomenon that garnered him a kind of celebrity status among Ufologists.

His initial fascination with the Phenomenon began about the time the Hills were undergoing hypnosis sessions to recover what happened during the period of "missing time" after which the Hills found themselves 35 miles farther along the highway that September evening. It was August 1964 and Hopkins and two of his friends saw a UFO during broad daylight over the Cape Cod town of Truro, Massachusetts. In a pattern familiar to researchers, this sighting was to inspire a lifelong interest in UFOs and related phenomena. In 1976 Hopkins published an article in the *Village Voice* concerning a UFO incident in North Bergen, New Jersey, that he investigated as a member of NICAP. That article generated a lot of interest from readers, and soon Hopkins found himself at the center of the alien abduction controversy. His first book on the subject—*Missing Time* (1981)—helped create the alien abduction "meme" and positioned Hopkins as the expert in the field. His interviews with many abduction survivors over the years contributed to an overall view of the Phenomenon as real, as remarkably consistent from one account to another, and as representing genuine contact between human beings and entities from elsewhere who were dispassionately examining humans for reasons of their own. He expanded on this theme in *Intruders* (1987) and *Witnessed* (1996).

It was in 1987, however, that the idea of alien abduction went mainstream. That was when Whitley Strieber's book—*Communion*—hit the bookstores and eventually became a feature film (1989) starring Christopher Walken as Strieber. Strieber was a screenwriter and novelist at the time, having written the horror novels *The Wolfen* (1978) and *The Hunger* (1981), both of which were made into feature films. In October and December of 1985 Strieber had an experience that changed his life forever.

It happened up at his cabin in the woods in upstate New York. He was awakened by a noise or disturbance on the ground floor, even though he had turned on the burglar alarm. He began to feel fear—the usual initial emotional reaction common to many abduction reports—and then noticed a small figure near his bedroom door.

The figure—which had the familiar appearance of what abductees and experiencers referred to as the "Small Grays"—rushed at him and Strieber lost consciousness for a while, recovering at a time when he had the sensation of movement, as if he was being carried. He was paralyzed, unable to move (which gave some observers the impression that he was experiencing sleep paralysis; however, the additional details of the experience argue strongly against typical sleep paralysis), and eventually wound up being held in a small chamber filled with tiny people. Various things were done to him, including a hair-thin needle inserted into his cranium and an unpleasant experience involving a rectal probe.

A succubus in the saucer.

He saw different types of beings, some squat and wearing blue coveralls, and at least one that he felt was feminine. His emotional responses ranged from fear, to absolute terror, to anger.

And then everything abruptly ended and he next regained consciousness in his bedroom at the cabin with only a vague feeling that something was terribly wrong. His emotional state went into a sharp decline, similar to what happened to Barney Hill twenty years earlier. Strieber's wife, Anne, reminded him that they had gone through a rough patch a few months earlier like the one he was going through now, and she was worried.

Eventually, the events of October 1985 and December 1985 would be linked to a series of alien abduction experiences occurring during that time. Strieber made contact with Budd Hopkins, who recommended therapists who were professional hypnotists, but Strieber selected his own so that there would be no chance of the therapist contaminating the recovered memories. The transcripts of those taped sessions comprise one of the chapters of Strieber's *Communion* so that readers can judge for themselves whether the therapist was asking leading questions or trying to guide the memory recovery in a specific direction.

As Whitley Strieber continued his investigation, more and more details began to emerge, not only of the specific abduction events of 1985 but of strange experiences from previous years as well. Strieber has discussed with us (Levenda) the possibility that the memories he has are screen memories designed to conceal or protect his consciousness from some awful reality. He has never, in fact, insisted that his experiences involved actual aliens from some other *planet*; only that the experiences are real and that the intelligences who populate the memories of these experiences are real. We explored the idea that some of his memories may be linked to the infamous mind-control experiments of the 1950s and 1960s. The fact that the young Whitley Strieber lived near Randolph Air Force Base in San Antonio, Texas—site of one of the largest contingents of Nazi scientists brought in under Operation Paperclip, scientists whose specialties ranged from aviation medicine to psychology—gave rise to speculation that some of what he had experienced as a child and, later, as an adult has its roots in experimental projects undertaken at that base at that time.

This relationship between the UFO Phenomenon and consciousness or "mind control" projects will be discussed and developed more fully as we go along. For now, however, it is enough to note that Strieber does take this aspect of the experience very seriously. That there is a heavy "consciousness aspect" to the UFO experience is undeniable. In fact, it would seem that many of those who later claimed an abduction experience were UFO witnesses first, sometimes on the very same day, as in the case of the Hills. This is why it is so difficult to separate the observation of what appear to be very material, very physical aerial craft from the psychological effects they seem to produce in witnesses. This is a unique aspect of the Phenomenon, something that sets it apart from other types of human experience. One can become traumatized by war or by violent crime, but how do we explain a traumatic reaction to the mere observation of a disk flying in the sky? Yet this seems to be what occurs on a fairly regular basis.

Strieber has written several more books that deal with his experiences and his interpretation of them. Strieber is a Catholic, so some of the titles of his works—*Communion, Confirmation*—have a distinctly sacramental feel. But he also has a background in Gurdjieff and Ouspensky, and at one point even investigated the Process Church of the Final Judgment at its headquarters in Mayfair, London, during 1968; that event was discussed during one of our interviews.

Sincerity is no determinant of truth, but it does argue against conscious deception. Strieber is sincere in what he reports, almost painfully so. In 2016, his collaboration with esteemed professor of religious studies Jeffrey Kripal was published as *The Super Natural*, a book that is groundbreaking in its approach to the Phenomenon. Here we have a specialist in religion and spirituality in conversation with an admitted alien abductee (perhaps the most famous living abductee) as they both try to understand the experience, where it comes from, and how to deal with it within an academic context as well as within a purely spiritual or psycho-spiritual context. Until that point, however, Strieber had been subjected to the usual ridicule and outright hostility of a broad spectrum of the population who jumped to all the usual conclusions before actually reading his work or making an attempt to understand it.

▼   ▼   ▼

That was where we stood on the question of alien abduction for another decade. Then along came John Mack, who directed his attention to the survivors of alien abductions and found many of the same themes occurring in their accounts.

Dr. John E. Mack (1929–2004) was a professor of psychiatry at Harvard Medical School and the author of several volumes in his field before writing a Pulitzer Prize–winning biography of T. E. Lawrence, *A Prince of Our Disorder* (1977). But in 1994 he published a book that would put him at the center of the Ufological controversy, *Abduction: Human Encounters with Aliens*.

He became intrigued by alien abduction victims because he thought the study of them would result in a diagnosis of a unique mental disorder. However, on the advice of his friend Thomas Kuhn—who authored the enormously influential *The Structure of Scientific Revolutions* (1962)— he approached the problem with an open mind:

> He told me not to worry about science and to watch out for the traps of language: real/unreal, inside/outside, psychological/external, happened/ didn't happen.[7]

This proved to be valuable advice and, in fact, the approach we have tried to take with this project as well. Mack realized that the people he interviewed—in the neighborhood of some 200 individuals—were suffering from symptoms of post-traumatic stress disorder but were otherwise psychologically healthy. Where he expected to find disability and delusion he found people who were intelligent, well-adjusted, even skeptical, but who all shared an experience that could not be explained within the confines of Western science.

In other words, they had experienced *something*.

Mack was criticized for *Abduction* and specifically for his conclusion that the abductees were discussing real events that they had experienced. Harvard University even held an investigation to determine if Mack should lose his tenured position and be fired. Although he had never said that the abductees had experienced real alien abduction or made a claim as to the "reality" of UFOs, etc., it was assumed that he had. Oddly, nuance seems to escape many in academia, and no less so in mass media. Critics jumped to the conclusion that Mack was saying that alien abductions were "real" in the sense that the abductees claimed they were. He had not claimed that at all, but as the quotation at the beginning of this chapter demonstrates, he did admit that a real Phenomenon was at play and that he had no idea what it was. That honesty was too much for many to appreciate. Mack managed to hold on to his tenure after celebrated attorney Daniel Sheehan (of Iran-Contra fame) and legal scholar (and fellow Harvard alumnus) Alan Dershowitz questioned the legality of the proceedings.

Nonetheless, the die had been cast. Mack, as a fully accredited member of the academic establishment, was now the most visible defender of the Phenomenon. Like Dr. James McDonald[8] before him, he had to deal with pushback from his colleagues and the potential of a ruined reputation just for taking the Phenomenon seriously and also taking the people who experience it seriously.

John Mack died in a traffic accident in England, hit by a drunk driver in London on September 27, 2004. Many conspiracy theories were developed over this event, but it does seem that his death was purely the result of an accident and not a deliberate attack. He was 74 years old.

Mack's contribution to the dialogue surrounding alien abduction experiences was on the same level as Jacques Vallée's decades-long contributions on UFO sightings and contactee reports. Dr. Thomas E. Bullard—a professor of folklore at Indiana University and a member of several UFO organizations—made a similar contribution to the literature, taking abductee reports as seriously as John Mack did and publishing *The Myth and Mystery of UFOs* (2010). This trend would continue to the present day with the groundbreaking work by religious studies scholar Dr. Jeffrey Kripal (Rice University) in collaboration with the world's most famous alien abductee, Whitley Strieber, in *The Super Natural* (2016), as previously discussed.

So what has been going on? Is there a context for the alien abductee experience, and to what extent is it related to the UFO Phenomenon, if at all?

▼   ▼   ▼

Between the sighting of a UFO and the abduction experience, there is an intermediary type of encounter. It's not an abduction, but it does leave physical traces on the experiencers, as do some abductions. This refers to encounters of human beings with the craft themselves, rather than with the occupants or presumed pilots of the craft.

Briefly, we will look at two such cases and the reasons why they are so important to a scientific study of the problem.

▼   ▼   ▼

What is probably not well known outside of a small circle of specialists is that there are scientists with impeccable pedigrees who have been studying the UFO Phenomenon in earnest, and most specifically the way that a close encounter affects the neurobiology of the contactee. Gradually, this information has come to light in books like Annie Jacobsen's *Phenomena*. Some of these scientists are either members of the To The Stars Academy founded by Tom DeLonge, or are colleagues.

Their area of study has centered on the physical effects produced by close proximity to a UFO, and their work has resulted in one of the first successful lawsuits brought by a contactee against the US military for injuries sustained during one such encounter. This is an important development and one that invites closer scrutiny. It is a Catch-22 situation for the military: in order to avoid civil liability for injuries caused by a UFO, the military would have to concede that the damage was not done by one of their aircraft or weapons systems, but by some other "unknown" aircraft or weapons platform over US airspace; conversely, if the damage was identified as having been occasioned by a military mission in the line of duty, the circumstances of that encounter would have to be described and identified in court documents, which then would demonstrate that while the military might be legally liable for the injury because the soldier in question was injured while on a military mission, the actual cause of the injury was a UFO.

One such case was the Rendlesham Forest episode, which involved US military personnel on active duty. The second was the Cash-Landrum affair, in which the victims were civilians. Oddly, these cases occurred within a day of each other in December 1980. The Rendlesham Forest incident took place at a USAF military base in England on the 26th and 28th of December that year; Cash-Landrum occurred outside a small town in Texas on December 29. In both cases, unidentified aerial phenomena were reported, and in both cases witnesses were injured as a result of proximity to the phenomena.

The Rendlesham Forest incident is controversial to this day. Numerous attempts have been made by professional skeptics to debunk the case—declaring that what US servicemen saw and experienced at the site could be everything from stars and planets to a lighthouse beacon—and we won't go into all of the details and arguments in this place. What is certain, and a matter now of public record, is that one of the US servicemen involved—Airman First Class John Burroughs—was injured, and his medical costs were eventually covered by the government in a tacit admission that his injuries were sustained as a result of his interaction with a phenomenon for which the military has no explanation.

The bare facts of the case are that on the night of December 26, 1980, what appeared at first to be either a fireball or a downed aircraft was spotted by a security patrol in Rendlesham Forest near the east gate of Royal Air Force (RAF) Woodbridge, a base that was being used by the United States Air Force at the time. John Burroughs provided an account of the sighting, in which he and two fellow servicemen followed what they thought was a light from a downed aircraft. There was a lighthouse with a bright beacon in the vicinity, and this was the fact that was seized upon by the skeptics who suggested that *all* the lights seen by the servicemen at Rendlesham Forest were explicable as a lighthouse beacon or as lights from a neighboring farmhouse, etc. That explanation does not account for the reason why the servicemen were scouring the forest for what they thought might be a downed aircraft in the first place, nor does the existence of other lights in the vicinity account for all the lights that were observed at the time.

(For instance, should a UFO land in New York City, one could account for some of the lights as coming from virtually any building in the city, but that would not account for the sighting of a UFO in the first place.)

Burroughs was told to wait a little distance from what they believed to be the crash site and act as a radio relay back to base, but fear and curiosity overtook him and he followed the two other airmen, Penniston and Cabansag, who went deeper into the forest to investigate. Penniston saw what he assumed to be a craft sitting in the woods, approached it, and actually touched the surface of it. Burroughs saw it and made sketches of it.

It was this proximity to the object that caused a rare medical condition in Burroughs.

According to Dr. Christopher "Kit" Green—a multi-credentialed medical doctor, scientist, specialist in psychiatry and radiology as well as in behavioral and neurosciences (including remote viewing), former analyst for the Central Intelligence Agency, and Assistant National Intelligence Officer for Science and Technology, among many other positions—John Burroughs was suffering the effects of something called broad-band non-ionizing electromagnetic radiation, such as that found in radio waves, thermal radiation, and light from different ends of the spectrum, such as infrared and ultraviolet. This type of radiation can cause burns and inflammation in living tissue.

According to statements made by Dr. Green, the government's medical files on Burroughs were classified for a long period. Even though Green himself has had Top

Secret TS/SCI clearances for most of his life, he could not gain access to Burroughs' medical file. This is very unusual, and again according to Green, the only medical files he knows of that were ever classified that high were the autopsy of President John F. Kennedy and the medical records of Adolf Hitler, putting AFC John Burroughs in very rare company. Reasons given included the fact that Burroughs' file contained multiple references to Special Access Programs and other matters that similarly were classified, making the declassification process a rat's nest of interlocking classification protocols. With a major push by several Senators, however, the files (or their salient details) were eventually released.

The story they told, combined with Green's analysis of Burroughs' symptoms, including damage to his heart, was astonishing. It offered clear evidence that something very unusual had happened to Burroughs and, as Green's investigation into this type of condition in other patients continued, seemed consistent with injuries suffered by other military personnel for whom there had been no previous diagnosis that made any sense. Partnering with Dr. Garry Nolan—a world-famous specialist in genetics research, and today a member of Tom DeLonge's team at To The Stars Academy—the two men came to agree that these patients had been exposed to electromagnetic radiation.

This and related cases suggested to Green and Nolan that contact with the Phenomenon could leave traces in human experiencers and contactees: biomarkers, including neurological and cardiological effects, skin lesions, some types of cancers, and even changes in DNA. This is an astonishing claim, but one that Green, Nolan, and their colleagues make quite soberly. Going where the evidence leads them, they assert that Burroughs' injuries were due to a damaged heart valve, which resulted from having been exposed to radiation from the UFO/UAP for an extended period of time and within close proximity. The Department of Defense and the Veterans Administration eventually agreed that Airman Burroughs had suffered these injuries as a result of the Rendlesham Forest incident during which Burroughs was on active duty, and thus he was eligible for medical benefits.

The victims in the Cash-Landrum affair—which as mentioned earlier occurred only a day or two after the Rendlesham Forest incident—were not so lucky, but the event is still important because it forced the US military into the position of having to say that whatever caused the medical problems in the three Cash-Landrum witnesses was not due to any aircraft owned or operated by any of the branches of the Defense Department; that it was, in fact, unidentified.

What is known about the latter incident is that Betty Cash, Vicky Landrum, and their little seven-year-old grandson Colby Landrum were in a car driving along Highway FM 1485 near the town of Huffman outside of Houston, Texas, around 9 pm when they saw an amazing sight. In fact, it was so amazing that the three eyewitnesses gave conflicting testimonies of what they saw. That something strange did appear in the sky over Texas that night was corroborated by other witnesses in the area, but none could agree on the contours or exact nature of the device, which seemed to be shaped like an upright diamond. What was agreed upon was that the bright light shone from a craft of some kind that sailed down from the sky and then hovered over the landscape at treetop level. A few minutes later, the device was surrounded by what appeared to be a fleet of Chinooks, the giant helicopters with twin rotors that are often used to haul troops and equipment. In this case, the witnesses were precise in their count of the number of Chinook-like objects encircling—at a distance—the strange craft. There were twenty-three.

Later statements by the various branches of the military known to have Chinooks in their inventory were clear: the helicopters had nothing to do with them, and they had nothing like the diamond-shaped craft in their possession either, leaving observers to suggest that there were only two remaining possible explanations. Either what looked like Chinooks were being disavowed due to the fact that their mission was highly classified and was connected with a UFO, or that they were alien craft, too! Since the latter was not very plausible, only the former made any sense at all: the helicopters were there to escort (or to maintain contact with) a craft that was somehow going rogue.

The light and heat emitted by the craft was intense, accompanied by a very loud roaring sound, accompanied at times by an erratic buzzing noise. Flames shot out of the bottom of the craft. The entire scene was so incredible that the three persons in the car felt that what they were witnessing was a religious event of some kind. Betty Cash and Vicky Landrum were devout, born-again Christians, and they interpreted the image of the descending light, the flame, and the roaring sound in typically Biblical terms, believing it at first to be connected with the Second Coming of Jesus and the End Times.

(It is worthwhile to note here that the case we made in *Sekret Machines: Gods* of human interpretation of UFO contact as communication with the Divine is reinforced by the automatic response of present-day pious Christians to a similar scene, as if revisiting the most ancient points of contact experienced by the Egyptians, Sumerians, Chinese, Indians, and others.)

The three got out of their car to see the craft more closely, but Colby Landrum was frightened and wanted to stay in the relative safety of the car. The damage, however, had been done. All three had been exposed to whatever was emanating or radiating outward from the craft. Even the surface of their car became extremely hot to the touch.

It was then that the Chinook-like helicopters showed up and hovered in a kind of formation near the craft. The flames shooting out from the bottom of the craft increased in intensity as it began to rise and move away, with the helicopters following it. The entire episode took about twenty minutes.

The three returned home, but there were other witnesses (including a police officer and his wife), and they confirmed the story about the fleet of helicopters. The latter was an important element, for it suggested US military involvement in the episode. But Betty Cash and Vicky Landrum almost immediately began to develop serious physical symptoms, including nausea, diarrhea, weird growths on their scalps and elsewhere, and burns: what appeared to be radiation sickness.[9] Their symptoms multiplied and got worse, and eventually Betty Cash and Vicky Landrum lost their jobs because they were unable to work and their medical bills were becoming astronomical. Betty Cash got the worst of it, as she was the one who stood in front of her car and was closest to the source of the flames. Alopecia also set in: their hair started falling out.[10]

The Cash-Landrum case, as it became known, resulted in their $20 million lawsuit against the US Government for damages that led to their illness being dismissed because the witnesses could not prove that the device that caused their injuries was in fact made and/or operated by the US Government. Because the various branches of military service all denied that the helicopters were theirs— and investigation showed that they belonged to no other organization or corporate entity—and they professed no knowledge of the diamond-shaped craft at the center of the affair, the episode therefore could not have been the fault of the government or the military. Thus the radiation sickness suffered by all three of the Cash-Landrum victims was caused by an unknown craft. In other words, by a UFO.

In a stunning presentation before a conference in Las Vegas on June 8, 2018, Dr. Hal Puthoff (laser physicist, specialist in space propulsion and alternate energy research, adviser to NASA and other government agencies for more than fifty years, one of the pioneers in the field of remote viewing, and a vice president of the To The Stars Academy) actually explained how this type of radiation sickness results from an anomalous vehicle—i.e., a UFO—"up-shifting" from infrared light to visible light in a specially engineered vacuum:

> So, in fact the infrared that you don't ordinarily see can get blue shifted up into the visible, so it's not surprising that all these craft should be so luminous. Now the downside from all of this is the fact that visible light, which doesn't have any particularly harmful effects, gets blue shifted up into the ultraviolet, so if you get too close to a landed craft you might get a sunburn, or off into the soft X-ray regions, so there's a chance of radiation poisoning.
>
> If you run across one of these sitting on the ground and it's powered up, I recommend you don't rush up.[11]

The biological markers that researchers like Dr. Garry Nolan, Dr. Kit Green, Dr. Hal Puthoff, and others have been studying are tangible evidence of contact between human beings and the Phenomenon, and the biological, neurological, and genetic traces left by this contact may tell us a great deal about the Phenomenon that even videos of aerial pursuits and gun camera footage may not. It's the "Man" variable in *Sekret Machines*; it's the Phenomenon leaving fingerprints at the scene of the crime.

OPENING THE DOOR

I broke open the door of the house of life, without knowing or caring what might pass forth or enter in. . . . I played with energies which I did not understand, and you have seen the ending of it.

– Arthur Machen, *The Great God Pan*

T HE 1970S SAW AN INCREASE IN ATTENTION BEING given to various aspects of the Phenomenon, from UFOs to spirit possession to science fiction and fantasy. In the popular culture we had the films *E. T.* and *Close Encounters of the Third Kind* (the latter boasting prominent French astrophysicist and Ufologist Jacques Vallée as a consultant). We also had the first installment of the *Star Wars* saga. We were being told that the alien presence on Earth actually may be benign, from the cuddly alien in *E. T.* to the vastly intelligent, if not a little manipulative, aliens from *Close Encounters.* The aliens in these films may not have been purely angelic, but they were far from evil and hardly threatening.

However, we were still reeling from the ideas presented in another genre of films. Movies like *The Exorcist* (based on the best-selling book by William Peter Blatty and "based on a true story") and *The Omen* were telling us that our children were either being possessed by demonic forces or were vehicles for the Devil himself. Aliens were non-denominational, but demons were all Christian. In fact, they were Roman Catholic! The *Rituale Romanum* (the basic ritual text of the Catholic Church) was used to exorcise the little bastards regardless of the human victim's own religion or lack of same. *"The power of Christ compels you!"*

It was not so much a factor of demons being Christian or an invention of Christianity as it was an extension of Christian power over beings that had once been gods or paranormal presences in pre-Christian, pre-Abrahamic, times. The demon in *The Exorcist* famously was identified with Pazuzu, an ancient Babylonian demon who was held responsible for famine and plagues of locusts but also was used to defend a home against attacks of the *lamaštu*: female demons who preyed on pregnant women and newborn babies. In fact, images of Pazuzu from the first millennium BCE are remarkably similar to those of demons in the popular imagination: a hideous expression, wings, talons, the tail of a scorpion, and even a serpent for a penis.

Some of this mid- to late-twentieth-century anxiety over aliens and angels may be due to the fact that the Vietnam War was coming to an end, and with it the counterculture optimism of the Sixties. The Watergate scandal had just begun, and the combination of these two ideas—political demons and demonic politicians—led some to believe that our leaders could not be trusted. To some, it also implied that our children—the Flower Power hippies of Woodstock who morphed into the violent bikers of Altamont—were evil. They were the enemy, conduits for the Devil's sinister plans. The ignominious end of the Vietnam conflict, with its searing image of desperate people clinging to helicopters leaving the roof of the US Embassy in Saigon, signaled the possibility of a weakness in the American psyche, some heretofore undetected flaw in the American dream. Nixon and Kissinger had promised Americans "peace with honor." What they got instead was the fall of Saigon and the killing fields of Cambodia. This catastrophic failure of military interventionism in Southeast Asia would be reprised in the equally disastrous campaign in Iraq more than twenty-five years later.

By the 1980s, the continued public interest (and official US government disinterest) in the UFO Phenomenon was further inflamed by a new twist. The cattle mutilations of the American West and reports of alien abductions everywhere began to shift perception of aliens as feel-good older brothers dedicated to our survival to amoral sociopaths tinkering with our DNA, implanting tracking devices, and . . . well, *probing*. The sexual deviance our ancestors expected from congress with the Devil—including the infamous *succubi* and *incubi*, demons who (respectively) visited unsuspecting men in their sleep and stole their semen, then impregnated women with the stolen semen and in the process created "hybrids"—now manifested in a most unexpected form: those that the most famous alien abductee, Whitley Strieber, calls the "Visitors."

How much of this was due to some kind of psychological dysfunction in the American psyche, and how much had a more tangible origin in the UFO Phenomenon itself? The problem became more complex, since reasonable people were ready to accept the reality of UFOs based on the preponderance of evidence—a mountain of it growing every day—but balked at accepting the idea that "space aliens" or "little green men" were abducting human beings in the middle of the night and experimenting on them. The whole alien abduction scenario was poised to submerge the entire UFO field by associating it with claims that were much more difficult to swallow than was the possibility of advanced aviation technology in the sky. Those who insist on a real-world explanation of UFOs—even extending to the extraterrestrial hypothesis that they are craft piloted by beings from another planet—have a hard time dealing with its corollary: that these same beings land, walk through walls, abduct human beings from their beds, experiment upon them, abuse them sexually, implant devices in their bodies, and return them to their homes only to abduct them again at some random time in the future.

Then there were those who insisted that the two phenomena—UFOs and alien abductions—were not related at all. The alien abduction phenomenon was linked to religious hysteria, shamanic-type experiences, or simply to psychological disorders, and therefore had nothing to do with flying saucers. The scenarios that some abductees reported of aliens, spaceships, and the like was considered to be culturally *inspired* rather than an expression of what actually *transpired*. In other words, today it's aliens in flying saucers, but yesterday it was witches on broomsticks. There is a core experience among humans that is reported differently depending on the era and the culture within which it takes place.

And that is exactly what we have been saying in this project.

The experience of religious figures, gods, angels, demons, jinn, may be expressions of contact with the Visitors. Therefore, alien abduction and experiences of contact with demonic forces are cognates: they are referring to a singular experience that is described differently by different people from different backgrounds. Therefore UFOs and alien abductions are related. They may be a step or two removed from each other (we don't understand the mechanism yet) but these "otherworldly" experiences derive from the same source.

As humans, we perform certain mental functions of which we are usually unaware. Our mind—our consciousness—sets up relationships between phenomena that might not exist in any kind of Newtonian context. We manufacture meaning and connections. Carl G. Jung called synchronicity (coincidence) an "acausal connecting principle." That pretty much sums up how most of us understand the world we live in.

The experience of the UFO is one such example of how we think and interpret our reality. Basically speaking, a UFO is nothing more than a machine that appears and disappears according to laws and motivations that are unknown to us. A *sekret* machine. There is no reason to assume that every UFO we see is actually piloted by some being inside of it. Many of them, perhaps most of them, may be completely empty. They may not even be material the way we understand the term. Some are, of course, if eyewitnesses are to be believed and there are enough solid cases to enable us to state this with confidence. Some may be the equivalent of unmanned aerial vehicles (UAVs), or drones. Some may be holographic images that are projected onto our world by a device unknown to us. There are many possible explanations, but few of them require beings that are using the UFOs as transport. Yet, because we associate flying machines with human operators, we assume that all UFOs have operators, albeit "alien" operators. We assume the UFO is a transportation device like an airplane or a helicopter, and not something completely different. These are logical conclusions, but they're based on too many assumptions rather than on the evidence. Since the civilian public is largely denied access to the evidence, the basis upon which their conclusions are drawn is reduced to speculation and assumption. Then an actual experience happens to one or more of them and—absent any academic, scientific, governmental, or military authority—they are forced to interpret it themselves.

When that happens, interpretation becomes representation; speculation becomes revelation.

This relinquishing of authority on behalf of trusted institutions may serve a short-term goal (kicking the proverbial can down the road, leaving the problem to some other group or organization to deal with at some later date), but in the long term it has eroded public confidence in the ability of the state or its component parts to inform, protect, and defend them. Prior to the December 2017 revelations concerning existence of the Pentagon's Advanced Aerospace Threat Identification Program (AATIP; the special Pentagon program that studies UFOs), the state had been able to contend with this situation through its ridicule of Ufology or by issuing blanket denials. Now, however, a tipping point has been reached, and this strategy is no longer viable. Too many individuals in government and industry with specific knowledge of the Phenomenon have come forward in recent years, and especially since 2015. In addition, too many individuals have been exposed to the Phenomenon and, in the Internet age, they are communicating with each other and forming their own opinions and strategies.

We've been here before. In the European Middle Ages, the response of the state (and the church) may have had more to do with ignorance of the nature of the Phenomenon than with an articulated policy mixing denial with ridicule. In fact, the general response was one of panic, hysteria, and fear. There was a paradigm within which to interpret the Phenomenon, however misguided, and it was Us versus Them, or more correctly God (and his church, and his state) versus the Devil (which included anything, any phenomena, which could not be explained or measured within that paradigm). Individuals were arrested, tried, tortured, and in many cases put to death on charges of having contact with the Devil. In addition, the material possessions of the accused were seized, which revealed a monetary aspect to the Inquisition and exposed a cynical dimension that had nothing to do with the spiritual war of good versus evil (or human versus alien) but rather an economic war of the state versus the people.

This is something to keep in mind when we examine the political implications of the Phenomenon. To study its history as purely one of technology, or of spirituality and consciousness, or of military-type conflict is to ignore its complexity; not only the complexity of its intrinsic nature, but also the complexity of its interrelationship with human institutions.

Over the years the US Air Force has stated that UFOs pose no threat to national security, which is why no resources have been devoted to tracking them, analyzing sightings, etc. Yet there have been many reports of UFOs harassing nuclear installations on the one hand or terrifying the civilian population on the other (alien abduction scenarios, radiation-type burns, and so on). In one case, in the Soviet Union, a missile installation was so badly compromised that it could have led to World War III.[12] In another case, in the United States, the nuclear missiles at one base were taken off-line at the time of a UFO overflight.[13] It would seem that these cases would merit a "national security" concern, but until recently there had been no evidence forthcoming from the government or the military of either country that such was the case. Until the December 2017 announcement in the mainstream media of the existence of AATIP there had been no acknowledgment that "Yes, this is a dangerous phenomenon with the capability of instigating global nuclear conflict; we have to begin a serious study with the goal of protecting our citizens."

In the case of the witchcraft hysteria in Europe during the Middle Ages, witches were considered a threat to the state as well as to the church. This is why instruments of the state were employed during the Inquisition to identify, arrest, imprison, torture, and execute those suspected of trafficking with the Devil. "For rebellion is as the sin of witchcraft" (1 Samuel 15:23): i.e., rebellion against the moral authority of the church (witchcraft) was identified with rebellion against the state (treason). Thus, contact with alien powers did have a national security implication that was consistent with the Catholic worldview that predominated in Europe at that time. Secular leaders derived their moral authority to rule from the church, and the church was the institution that determined the legitimacy of paranormal phenomena. It also determined what was, and was not, acceptable in the scientific realm (as seen in the trials of Galileo, Giordano Bruno, etc.). So the Church was able to say that paranormal phenomena existed (making an essentially scientific statement concerning the reality of phenomena experienced by the people) but that they were the work of the Devil (a spiritual judgment, warning people of the dangers of trafficking with the phenomena as well as the punishments that would ensue).

It is important to realize that "reality" was a fluid concept before the scientific revolution arrogated to scientists the final word. Reality was composed of both natural and spiritual phenomena, with religious concepts at the heart of the worldview. The very word *real* is problematic, for it derives from an Indo-European root that has given Romance languages, for instance Spanish, the word *real* as in "royal" as well as *real* for "reality," thus leading us to understand that "reality" was whatever the "royal" said it was. Reality, in this case, is not what we experience; reality is a social construct of science, technology, and cultural attitudes that has been crafted by specialists who are in the employ—via government grants—of the state, i.e., of the "royalty," and which can change considerably from culture to culture, society to society.

Civilian authorities realize, however, that personal experience can be in conflict with—or entirely opposed to—what the state considers "reality." That is where the relatively new discipline of psychology comes in. Psychology picks up where religion left off: it fills a vacuum created by the scientific revolution. Psychologists can point to behavioral motivations, unconscious states, instincts left over from an earlier, prehistoric age, etc., as the malleable and therefore faulty mechanisms that cause us to see things that are not there, in the sense that they are not part of the sanctioned reality of science and state. Love—which for Marsilio Ficino, Giordano Bruno, and so many others of the Renaissance and the centuries since then was a force of nature, a magical emotion that effected communication and influence across time and space, even "spooky action at a distance"—became something entirely personal, i.e., psychological, and incapable of being measured. Love was analyzed as being the result of a complex of hormones, pheromones, and deep-seated, unresolved psychological conflicts and neuroses, of which Freud's Oedipal and Electra complexes serve as examples. In their zeal to be accepted as scientists, since science is the only game in town, psychologists attempted to impose standards of measurement and predictability onto the human mind, resulting—after World War II—in the first-ever *Diagnostic and Statistical Manual of Mental Disorders* (DSM-1).

Many do not realize that the DSM was *not* the brainchild of psychologists doing pure science but was the result of a mandate from the *United States Department of Defense*. The idea behind the DSM was to discover a way to defeat attempts by soldiers who were seeking ways of escaping active duty by claiming shell shock or other traumatic conditions. The Army had no legal way of knowing if the soldiers were faking mental disorders, and they were in desperate need of having as many soldiers as possible return to the field. Thus the DSM was born, bringing a degree of respectability to the soft science of psychology—or at least acceptance of it by the state.

In the old days, paranormal phenomena could be addressed and identified by the church, which would then take appropriate measures to counteract their effects. There was a defense against demons and witches and their hybrid offspring: exorcisms, the Holy Mass or Divine Liturgy, prayers, holy water, the sanctuary of the church. Until recently, paranormal phenomena in the modern age had no defense; the split between the authority of the church and that of the state is pretty well defined, with the state having the upper hand through its surrogate, the scientific establishment. Thus, paranormal phenomena were deemed not to exist, to have no *reality*: what is called paranormal is the result of delusion, ignorance of scientific principles, psychopathology, or just simple credulity. Thus, there is no defense against it.

Or, to put it another way, since there is no defense against the paranormal—including all aspects of the UFO Phenomenon—then *it* does not exist. Problem solved.

While Europe in the fifteenth century might have been obsessed with ideas of succubi and incubi, witches, and Devil worship, other parts of the world believed that nonhuman beings existed at the fringes of "reality." Among the Arab populations of the Middle East and North Africa there was—and still is—a belief in the existence of the *jinn*. The jinn (from which we get the English word *genie*) are corporeal, they reproduce, they have free will, and they live much the way human beings do, but they are beings composed of a "smokeless fire." Some of the jinn even have converted to Islam, according to Islamic tradition. Indeed, some of the figures known as demons—such as Satan or *Shaytan* (known in Islam as *Iblis*)—are actually evil jinn. This not only shows a point of connection between Islamic tradition and Christian beliefs but also shares some common ground with alien abduction scenarios in which the aliens are perceived much the same way that Muslims perceive the evil jinn, as well as with the school of thought among some Evangelical Christian groups that aliens (as popularly understood) are actually demons.

But are "aliens" the same thing as "extraterrestrials," i.e., actual beings from another planet? A number of recent books have popularized the idea that aliens are extraterrestrial beings: grotesquely sinister creatures who are either hostile to humans or simply behave in ways that are dangerous to humans. The famous physicist Stephen Hawking even went so far as to warn us against commerce with extraterrestrials,[14] and for good reason. We have no way of knowing how an extraterrestrial race would interact with Earthly humanity. The possibilities for communication would appear to be nonexistent. A race evolving on another planet and subject to entirely different environmental conditions would not possess the same set of sensory instruments (ears, eyes, mouth) that we do, and could—if they existed at all—be composed entirely differently and would interpret different data and different stimuli. Their somatic structures would be so different from ours as to render any identification of humans as similar beings almost impossible. They might see us the way we see insects, or animals, or plants—if they see us at all. We have no idea if concepts like empathy or sympathy would exist among beings of another planet; if they don't, then we can't expect an extraterrestrial race to care about what they do to us.

This is the official, NASA-sponsored opinion as represented by a study they commissioned and published in 2014. Titled *Archaeology, Anthropology, and Interstellar Communication*, it was edited by Douglas A. Vakoch (director of interstellar message composition at the SETI Institute and a professor of clinical psychology at the California Institute of Integral Studies) and included essays by sixteen contributors on subjects ranging from SETI (the search for extraterrestrial intelligence) to the archaeology of the ancient Greeks and the Maya to ethnology, evolution, and culture. The consensus among the scholars consulted was that communication between the human race and an extraterrestrial race would be virtually impossible, due to the conditions mentioned previously.

This implies that even colonization would be impossible. A look at world history of the last few thousand years reveals that although a nation could colonize another on the other side of the world, a modicum of communication was necessary to do so. The Dutch may not have been fluent in the Indonesian languages when they arrived in Java, for instance, but they found there human beings like themselves with needs and desires so similar that learning their language was facilitated (and, of course, so was their colonization effort).

What is more likely in an ET visitation scenario is what happened when Europeans colonized the Americas.

In that case the Native Americans were considered to be nonhuman, at least by the English and many other European commentators. This was due to the fact that there was no mention of the Native American "race" in the Bible; hence, the "Indians" must be devils. This was an excuse to engage in one of the modern world's most infamous campaigns of genocide.

To the Spanish conquistadors, the Native Americans might have been human, but they surely weren't Christian. Thus, the indigenous inhabitants of Mexico and the lands south were forced to convert to Catholicism or die. It was not only necessary to accept the secular authority of the church, but to adopt its spiritual worldview: to abandon traditional culture, to take new names, to close the door on direct communication with the Other and to leave all of that to the priests.

Race and religion were used as rationales for the enslavement or annihilation of the Native American populations of both North and South America. The race issue was probably the worst-case scenario; although one could, theoretically, convert to another religion, one cannot "convert" to another race, much less to the human race from a race of demons. We have to assume, therefore, that an extraterrestrial race arriving on our planet could have the same attitude toward us. The popular literature that characterizes aliens as demons could as easily be understood in reverse: that human beings are the demons and the aliens are the real humans. Wouldn't an outside observer, coming to our planet for the first time and seeing how we slaughter each other and the planet with reckless abandon—like three-dimensional characters out of a painting by Hieronymus Bosch—come to that same conclusion?

Or would they see us as a race of termites, slowly destroying the home we live in and deserving only of extermination before the whole house comes down?

▼   ▼   ▼

In the end, the anthropologists hired by NASA to contribute to their study on alien-human interspecies communications entirely ignored an important and vocal demographic: those human beings who claim, year after year, that they have indeed communicated with aliens. In fact, literature of this type goes back thousands of years. What is required is a scientific appraisal of those communications—from the point of view that it represents something real and not the ignorant delusions of pre-technological peoples—in order to isolate some common denominators, not only in message but in the transmission of those messages, thus contributing to the overall discussion in ways that officialdom has yet to appreciate.

Thou Shalt Not Suffer a (Witch, Alien, Contactee) to Live

How did we used to handle contactee reports?

Beginning about the fifteenth century, European Catholics saw the dissemination of official information concerning the Devil in the so-called Witchcraft Manuals. This began with the *Malleus Maleficarum* ("Hammer of the Witches") of Heinrich Kramer and James Sprenger, published circa 1484. Kramer and Sprenger were Inquisitors, members of the Dominican order, and highly educated. They were famous as witch hunters, and in that capacity generated the *Malleus Maleficarum* for the use not only of Catholics but also Protestants, for witchcraft was of grave concern to everyone in Christendom.

To the Inquisition, there was a human agency at work in the phenomena they faced: dead and mutilated cattle, illnesses of unknown origin, the constant reports of succubi and incubi, and the Witches' Sabbat. To the Inquisitors, the human intermediary between the paranormal and polite society was the witch.

The witch could be male or female, but was assumed to be female in most cases. Witchcraft was used to explain all sorts of events that today would be considered as having natural causes: unwanted pregnancies, the plague, dead cows, spoiled milk. The fact that superstition can be used to explain away most of the reaction to these events does not address the underlying terror that witchcraft represented. Something was going on and, like today, perhaps 80 to 90 percent of cases can be explained by natural phenomena or other—nonparanormal—causes; but that leaves the 10 percent, or less, for which there was no rational explanation.

Today there is no one in the UFO community who would claim that sightings are the result of witchcraft, or that alien abduction scenarios reflect the Witches' Sabbat.[15] In other words, human agency in these events has been ruled out as a contributing factor. In fifteenth-century Europe, however, virtually all paranormal phenomena were believed to be the result of spiritual forces at work in the world through human agents: the witches. This resulted in a purge of those who were believed to be cooperating with these spiritual forces or who had some special contact with them.

The Witches' Sabbat is a perfect example of a fifteenth-century perspective on twentieth-century alien abduction. In the fifteenth-century version a human being—usually a woman, but sometimes a man—goes to sleep and wakes up at a great distance from her bedroom in a group of humans and demons, usually in a desolate area remote from human habitation. Sometimes there is only one demon—the Devil himself—who performs sexually with the victim (i.e., the witch) or who expects sexual favors from the victim. In many cases the victim has flown through the air to the Sabbat; in other cases she simply goes to sleep at home and wakes up at the Sabbat. At the end of the meeting, the woman either returns to her bed by flying through the air or simply wakes up the next morning.

What is generally forgotten about the "confessions" elicited from "witches" under torture is the fact that many of the self-confessed witches did not find the Sabbat pleasurable. In fact, they recoiled from the act of intercourse with the Devil for various reasons. In some cases—at least, according to the Witchcraft Manuals—they were made to kiss the Devil's buttocks as a sign of obedience and loyalty. In other cases they were penetrated by the Devil, whose member was said to be icy cold.

The benefits derived from being a witch were somewhat pathetic for all of that. One would expect worldly powers of a scale seen in the film *The Omen*: the ability to cause wars and global conflagrations. Instead, a witch usually was accused of poisoning a neighbor's cattle, causing the milk to spoil (evidently a major issue at the time, if the Manuals are to be believed), or of more dire actions such as causing infants to be stillborn or, if they were born healthy, of ensuring their deaths shortly thereafter. That is all they were able to achieve for their attendance at Sabbats and kissing the Devil's hindquarters, risking as well their eventual discovery by the Inquisition and certain torture and death.

Eventually it was realized that human agency did not play such an important role in the witchcraft phenomena. Even in the infamous case of the Salem, Massachusetts, witch trials (1692), in which dozens of persons, mostly women, were accused of witchcraft, and nineteen of them executed, the political authorities finally realized that they had had enough and called a halt to the entire proceedings. What is fascinating to learn about this episode is that Tituba—the West Indian slave who generally is blamed for having started the Salem witchcraft hysteria—was released from prison and went on to live a normal life after the trials and executions were over.[16] Even Tituba, with her tales of sorcery and spells and flying to Sabbats—and including a Man in Black(!)— was exonerated because it was believed that she had no influence over the paranormal events that took place at Salem (although no other explanation was given).

Today we have cattle mutilations that are inexplicable and are often thought to be the result of nonhuman interference (as the evidence clearly suggests, and as our own highly placed informants agree). We have alien abductions that share many elements in common with the traditional Witches' Sabbat, from flying through the air to sexual penetration with a cold instrument. The difference today is that we do not accuse human beings of *causing* these phenomena but of *being victimized* by them.

That approach might have something to do with politics.

> Witchcraft was inexplicably linked with politics.
> —Montague Summers[17]

The alliance between governments and science proved to be a two-edged sword. In the beginning of the scientific revolution secular governments could rely on the world of science to provide advanced weaponry as well as improved methods of navigation, manufacturing, metallurgy, medicine, and chemistry. While religious institutions might have been suspicious of science and scientists, the kings and nobles could not afford to ignore the tools—the technology—that were the fruit of scientific research and scrupulous adherence to the scientific method, which (ostensibly) did not allow of ideological influences; rather, it was based on evidence, on observable and measurable facts. That meant there was no room for witches and demons in the hallowed halls of rationality. The more the invisible forces were sidelined, explained away, or ridiculed, the less the church could enforce its authoritarian will over the people.

Gradually religion came to be located in a specific box, or category, of human experience with little or no contact with the other boxes. The worldview represented by the scientific revolution was that of the waking world; the world of sleep and dreams was surrendered easily to the church. Even philosophers were hard at work developing systems of thought and categories of knowledge that had nothing to do with the God of the Hebrews or the Jesus of the Christians, much less the pantheons of Africa, India, China, or the indigenous populations of the New World and the Old.

By the nineteenth century and into the twentieth, the church was under increasing attack from the scientists, or at least it seemed that way. Darwin's theories challenged the Biblical history of creation; archaeological digs in Egypt and the Middle East threatened Biblical authority over the history of ancient civilizations. People who were not priests and not even Catholic or any other kind of Christian were making statements about the world, about reality, and about human origins that were not sanctioned by spiritual authority. Attention was drawn to the "mysterious East" with a growing fad concerning all things "Oriental" and correspondingly non-Christian. Helena Blavatsky was in Central Asia and India; Alexandra David-Neel was trekking to Tibet. If the church and the Protestant Reformation had somehow failed to deliver the spiritual goods to Europeans (had failed to stand up to the scientists), then India, Tibet, and China were the last resorts, which implied (but actually never accomplished) an eventual jettisoning of Western values altogether.

Why didn't the world—at least, the Western world—simply surrender to the bland blandishments of the scientific worldview and abandon its attraction to spirituality when it was obvious that the scientists had the fancier toys and the stronger arguments? Why did Westerners want it both ways? Why this insistence on the mysterious, the esoteric, the occult, when science was shining a light into all the dark corners?

While historians have tried to give various explanations for the survival of spirituality (i.e., "superstition") well into the twenty-first century, and as atheists have published everything from irritated manifestos to entire volumes dedicated to the subject, one essential element of the problem is often overlooked: the close association between governments and scientific establishments replaced the close associations that had existed previously between governments and religious establishments.

The twentieth century saw the rise of a political system that took this position to its logical conclusion, one that was determinedly anti-religious and atheistic: communism. The Soviet Union officially downgraded the role that religion—in this case, the Russian Orthodox Church—would play in the new system. Membership in the Communist Party would be jeopardized by church affiliation. The same was true in Mao Ze Dong's China. Atheism was a prerequisite for Party membership, and Party membership was a prize to be won and cherished. Science, and especially technology, was elevated above everything else. In the communist system the state owns the means of production, whether factories or farmland, hammer or sickle. The church produces nothing, so it has no value for the state; participation in religious services is a waste of time and human capital, and thus of productivity. Church buildings were seized and turned into offices or museums. Church officials were murdered, or sentenced to hard labor, or pressed into service to the state.

One result of this policy was the loss of any mechanism through which to interpret or manage the Phenomenon. With the suppression of religion came the accompanying suppression of any constructive means of addressing what everyone knew to be true: that the Phenomenon exists, and that even in Mother Russia there were unidentified flying objects buzzing nuclear weapons silos and almost causing World War III.

In Western countries, of course, the same events were taking place, but because there was no censorship of the press, the reports of UFOs, cattle mutilations, and alien abductions would appear in both mainstream and "fringe" media. The scientists and skeptics in the West could blame gullible persons for these sightings and ridicule their insistence that the Phenomenon is real: a form of censorship by sarcasm. In communist countries, the problem was somewhat different.

Censorship meant that many of these reports could be suppressed by the state, and they were. It also implied, however, that what sightings *were* reported were the result of actual observation and could not be blamed on superstitious witnesses. After all, the Soviet Union was a country in which no superstition (officially) remained. That put Soviet science in the awkward position of having to explain the Phenomenon in scientific terms. Scientists permitted only two possible scenarios: mistakes by untrained observers who did not understand that what they were seeing were natural phenomena, or in some cases rocket launches for which there had been no advance announcements—this being the Soviet Union, after all—and which were mistaken for balls of light, etc. They had to entertain the "ETH" (extraterrestrial hypothesis) just for the sake of completeness, but they had no confidence in that explanation.

Like their American counterparts, they went to great lengths to prove that the vast majority of UFO sightings were explainable (and in most cases the result of secret missile tests). Of course, they were still left with a small percentage that remained unexplained, and the ratios are surprisingly similar between the US and the USSR. For both, more than 90 percent were explained as natural phenomena, weather balloons, astronomical events, and identifiable flying objects such as advanced aircraft or rocket launches, leaving 10 percent (or less) as genuine unidentified flying objects, though with a great deal of official certainty that they did not represent extraterrestrial spacecraft.

In fact, the Soviets did not even use the term or the acronym "UFO" due to its suspect American origin. Rather, they preferred the term "paranormal phenomena," a phrasing that suggests they situated UFOs within the broader field of "rejected knowledge." Published reports by Russian experts after the fall of the Soviet Union imply that the UFO Phenomenon in the Soviet Union was nowhere near as pervasive as that in the West (and especially the United States) and that they had no alien abduction cases at all. This led them to opine (humorously) that aliens obviously were not interested in kidnapping Russian citizens. The implication, however, was more serious: it was a critique of the entire field of Ufology as a peculiarly Western (and hence decadent) form of mental aberration. Reports of "flying saucers" were due to the credulity and weak-mindedness of Western peoples, who were known for their superstitious and unscientific attitudes.

Unfortunately, however, the hubris of the Soviet experts had little support in reality. The collective hallucinations of the Russian military and intelligence classes share a lot in common with those of their Western counterparts. There *were* alien abduction cases reported in the former Soviet Union. There *were* close encounters that would eventually be included in the voluminous files of Western UFO research groups. In fact, the presence of UFOs over Soviet military bases became so well known, and to an extent predictable, that methods were developed by the Soviet military to determine whether communication with these objects was possible. Thus, to insist that the Phenomenon is a culturally conditioned hoax or delusion is to ignore the evidence. *Interpretation* of the Phenomenon may very well be the result of prevailing cultural attitudes—either in purely "religious" terms or purely "scientific" terms or even some uneasy combination of the two—but the Phenomenon does exist and does challenge the resources of governments both capitalist and communist, democratic and totalitarian. And when unidentified aerial phenomena can be observed as radar traces, or when their presence coincides with advanced technology malfunctioning, then cultural attitudes—claims of religious mania, ignorance, gullibility, mass hallucination—are virtually worthless.

After all, machines do not hallucinate.

▼　　▼　　▼

There are those who wish to extricate the "alien" hypothesis and its associated reports from the purely mechanical aspect of the Phenomenon, insisting that two entirely different phenomena are involved. The former, they insist, has more in common with mythology, psychology, and religion than it does with the actual flying objects themselves. If only one could study and examine the UFOs without any of the messy associations of humanoids, or Small Grays or Tall Nordics, etc., then maybe some progress could be made. The craft are not necessarily piloted the way we understand it. They may not even be under conscious control. Why posit an alien sitting at the instrument panel? Who says there is anyone in there at all?

Part of the problem began with Dr. J. Allen Hynek. He proposed what would become the famous "close encounter" system.

A close encounter of the first kind was a simple sighting of a UFO at close range, up to 500 feet away. The machine itself.

A close encounter of the second kind was an effect the UFO seemed to have on electronic equipment or other traces such as electrical, chemical, physical, etc. Again, only the machine itself is the subject of this type of encounter.

A close encounter of the third kind, however, is when a creature is observed, such as a robot, a humanoid, etc. In other words, an "alien" of some kind. At this point the connection is made between the machine and an entity associated with the machine, implying conscious control of the Phenomenon. Thus, what Hynek has done is assume that the appearance of "aliens" and the observations of the machines themselves are linked. There is a kind of continuum of escalating revelation from seeing a machine or device of some kind unknown to human technology to the effects the machine's presence has on mechanical and biological material in its vicinity and then, suddenly, a leap to the infamous "little green men." One could argue that the "close encounter of the third kind" was totally unnecessary to Hynek's scheme; that those who saw aliens wandering around the vicinity of UFOs were of a different category of witness altogether than those who saw unidentified flying objects that left radar and other traces that demonstrated their objective reality.

In other words, are we projecting our own idea of "man and machine" onto a phenomenon that has nothing to do with how we understand machines? Do we see aliens because we need to see them, need to understand the Phenomenon as deliberate, consciously controlled, mechanically manufactured transportation devices for the beings who had built them?

Is the "alien" a liminal figure we create in order to get our minds around the fact that a phenomenon exists for which there is no other rational explanation? Have we anthropomorphized the Phenomenon in this way, super-imposing a quasi-human face on the mysterious—the *sekret*—machine?

Almost.

OF GOLEMS AND GRAYS

Among the better-known of these legends is the one connected with the name of Elijah of Chelm (middle sixteenth century). . . . He was reputed to have created a golem from clay by means of the *Sefer Yeẓirah*, inscribing the name of God upon its forehead, and thus giving it life, but withholding the power of speech.

– Joshua Trachtenberg[18]

AS WE DEMONSTRATED IN THE FIRST BOOK OF THIS series, sightings of flying objects in ancient history were always associated with "supernatural" beings of some kind, such as gods, demons, angels, etc. That association has been made since the time records were first kept. This is not a new development in the saga of the Phenomenon but a characterization of it that goes back in time to Ezekiel, the *Mahabharata*, the *Enuma Elish*: the flying objects are piloted. The pilots are not human, but they are almost human. They speak with us. They communicate knowledge and wisdom. Or they urge us to violence. Or they fill us with despair. It's possible that we have been projecting our own image onto the Phenomenon since time immemorial. That may be the key to understanding it.

In Genesis, we are told that God made human beings in his "image and likeness." From that perspective, the human race is a projection of the mind of God. *We* are the liminal figures. We exist in a plane that is partway between the divine and the earthly. The aliens, then, are wholly Other.

Consider this. According to Jewish and Christian tradition, angels do not *sit*. They stand eternally. In fact, *angels have no knees*. One of the earliest statements to that effect can be found in a midrash known as the Genesis Rabbah,[19] composed about 400 CE. This theme was taken up around the same time by Saint John Chrysostom in his homily based on 1 Corinthians.[20] What a strange image, the angels with no knees. This idea was reinforced by the *merkava* and *hekhalot* literature of early Jewish mysticism dating to about the same period. In this tradition, the mystic—who has been successful in rising up the seven heavens to reach the penultimate stage—sees Metatron sitting on a throne. Because angels do not sit, the untrained imagination would conclude that Metatron must be God. That mistake could cost the mystic his life; Metatron is not God (there are "no two powers in heaven") but rather the very human Enoch, the prophet who was taken up bodily into heaven. If one mistakes Metatron for God, one is slammed back down onto Earth, never to make the ascent again, for one dies as a result.

During the Divine Liturgy of the Russian Orthodox Church, one stands for the entire three-hour service. There are no pews, and only a few folding chairs around the walls for the elderly who are not able to stand for extended periods of time. Is this in emulation of the angels standing eternally around the throne of God who, presumably, does sit and does have "knees"?

Why is this important, you ask? Because *aliens* have no knees:

> The Small Gray's body appears frail, with thin limbs and no musculature or bone structure. There are no "knees" or "elbows" as such, and legs are the same diameter from the top of the thigh to the bottom of the calf. Nor are there clearly defined "ankles or wrists."[21]

That quote from Bryan fairly accurately reflects the way many experiencers describe the "Small Gray" type of alien. This is such a strange detail—yet reflective of ancient Middle Eastern ideas about angels—that it may be yet another piece of textual evidence of an actual observation that has been "verified" by more modern "mystics": the abductees. The information concerning the angel's lack of knees is not something that would be well known or discussed outside a small circle of religious studies scholars, yet the observations of nonspecialist citizens are seen here to validate a concept of which they previously knew nothing.

Indeed, one is hard-pressed to find examples in alien abduction literature of aliens who sit down. This pose of standing eternally was understood in Biblical times to mean that the angels were constantly singing the praises of God and would never presume to sit in front of Him even if they could. (One could say, half-humorously, that the angels' lack of knees was the product of a kind of celestial genetic selection.) This did not mean that angels could not walk; like the aliens, their gait would appear somewhat robotic to our eyes, a stiff-legged strut like Russian soldiers on parade. Luckily, they had wings for longer trips!

Angels would appear and disappear at will. They could pass through walls like the Small Grays, and had many other attributes in common with alien stereotypes. High intelligence, a kind of serenity born of a wisdom with which we humans have no experience, supernatural power, the ability to know the future, and most important, their true function—revealed by the Greek version of their name, *angelos*—of messenger. Like the aliens, angels arrive with important information, usually warnings of dire catastrophes to come. In some cases, albeit rarely, their presence also contributes to pregnancy, even of women normally considered beyond childbearing years. Mary, the mother of Jesus, is one such woman who was visited by an angel and who then became pregnant. Elizabeth, the mother of John the Baptist, conceived the same way, even though she was advanced in years (Luke 1: 13–25). The impregnation of women by aliens is a staple of UFO literature. Then, of course, we have the case of the "sons of God" and the "daughters of men" that we discussed at some length in the first book.

While numerous attempts have been made to draw comparisons between Biblical angels and the aliens of modern experience, the lack of knees in *both* species is normally not mentioned. It also should be noted that nowhere is there a suggestion that angels have reproductive organs; the same is true of the Grays:

> The lower part of their anatomy does not contain any stomach pouch, or genitals; it just comes to an end.[22]

If it is held that the angels are responsible in some way for the impregnation of both Mary and Elizabeth, then an alien abduction parallel is not too far-fetched. The aliens themselves do not seem to reproduce in the same fashion as humans or other mammals. Their impregnation of human females, therefore, must be accomplished in some other way. Hence the repeated accounts of sterile laboratories and the aliens "operating" on women, or taking semen from abducted human males and using it to fertilize the ova of abducted human females.

Which brings us back to the Middle Ages and the realization that there is a succubus in the flying saucer:

> Abductees see no eating quarters, sleeping quarters, no evidence of food or drink aboard the crafts. "What do we make of this? . . . A humanlike figure which under its skin is very, very different. They do not appear to breathe or ingest food or water."

Someone from the audience remarks, "*Everything you have described sounds more like machinery than biology . . .*"[23] (emphasis added)

Beings that come from an environment other than our own, who do not eat or drink, who are different from us in physical appearance, and onto whom we project supernatural characteristics can be angels or demons, gods or aliens, or even robots or androids, because we simply have no earthly frame of reference that makes any sense. The legend of Oannes—the supernatural being who came out of the sea in ancient Mesopotamia to instruct the people in various arts and sciences—describes a very similar concept.[24] Oannes did not eat or drink, and this peculiarity was noted by the chronicler. Oannes was dressed oddly and arrived from the depths of the sea in his own vehicle. Oannes expected to have the attention of the people he encountered; he expected them to listen to him. He knew that, for all his strangeness, his otherness, he would not be attacked (or perhaps he was not afraid of being attacked, knowing that the humans could not harm him). Abductees recall similar experiences. There is no resistance to the alien presence, perhaps due to being rendered helpless by alien technology. There is fear, however, even terror.

Communication takes place via images rather than language as we understand the term. This comes close to the thinking processes of those born deaf. Hearing people think in terms of spoken language, the "voices in your head." The deaf think in terms of sign language and images. In the absence of sound, images take on greater significance and form patterns of thought. The aliens as described by abductees communicate with humans the same way, which is consistent with what we "know" of alien biology—speaking here of the Small Grays—which does not appear to include ears or a mouth. If we are to believe the abductee accounts, the Grays not only think in terms of images but can "transmit" these images as a form of communication. We would consider this a form of mental telepathy, but it may be only the result of a communication method that evolved naturally among beings for whom speech (in our sense, as a function of sounds as well as meanings) was not possible.

The contactee experience often includes warnings and prophecies coming from the aliens. They do not communicate new technologies or instruct in arts or sciences. Like the experiences of Old Testament prophets whose "angels" brought only messages of dire consequences, the aliens seem to be worried or anxious about the state of affairs on Earth, perhaps mirroring humans' own concerns. While the experience of alien contact or alien abduction is unique and often characterized as "high strangeness," the information collected during these experiences is surprisingly pedestrian. Warnings of global catastrophes have been a staple of divine prophecies since the earliest Biblical writings in the West and have percolated into the popular culture (as, for instance, in both versions of the film *The Day the Earth Stood Still*).

There is another aspect of this experience that should not be overlooked: human beings are themselves treated as if they are machines.

It has been noted that at times experiencers feel that the alien presences are not organic beings like humans or other mammals. They are sometimes described as insectoid or reptilian, but at other times are compared to robots and mechanical devices bearing only a superficial resemblance to humans or humanoids.

However, these beings—however they are described—seem to treat human beings as objects. In some cases humans are poked, prodded, examined with a variety of mechanical or electrical-type instruments, and then sent back "into the field" to be picked up again for further study. This intrusion is not limited to the bodies of the humans they allegedly abduct, but includes their minds as well. It represents a manipulation of the entire being, which often results in the kind of fine-tuning that some abductees report. Human beings are treated like marionettes, as simulacra of autonomous beings, devices that need fixing or adjusting. It inspires a reversal of roles, making us the machines and the "aliens" the ones who are the actual autonomous beings, which is pretty much how the creation epics of many cultures describe how humans were made, and why.

The human being as golem, or homunculus.

In the Jewish legends surrounding the sixteenth-century Rabbi Judah Loew of Prague, the Golem is created to defend the Jewish ghetto from pogroms initiated by Emperor Rudolph II against them. The Golem—created out of clay like the first humans—is "activated" by the insertion of a parchment scroll in its head with the word of God written on it.[25]

We may consider this the first implant.

While there is doubt that Rabbi Loew was ever actually involved in creating a golem, there are legends even more ancient that do discuss the creation and management of golems.[26] According to the Talmud,[27] Adam himself was created as a golem: a tradition that is consistent with ancient Sumerian and Babylonian traditions. If humans are golems, then humans are machines: artificially created devices manufactured for a specific function. Strange, then, how the alien abduction scenario often includes accounts of implants. These are often imagined to be tracking devices, such as those used to tag animals, but that may be the result of projecting a concept onto the idea of the implant rather than an evaluation of the device itself. It's possible that aliens have no need of tracking devices. What if the purpose of these mysterious implants is something far more ambitious? What if the implant is used to "activate" a human being in some unexpected fashion?

In the 1960s tremendous strides were made in the development of electrical stimulation of the brain. This work—begun with Dr. John Lilly but continuing under the CIA's Office of Research and Development (ORD)—was designed to enable the remote control of animals and then humans via an electrode planted in the brain. This research was augmented by genetic engineering projects that were years ahead of work being done in the private sector. As investigator John Marks revealed as early as 1979:

> "We looked at the manipulation of genes," states one of the researchers. "We were interested in gene splintering. The rest of the world didn't ask until 1976 the type of questions we were facing in 1965. . . . Everybody was afraid of building the super-soldier who would take orders without questioning, like the kamikaze pilot. Creating a subservient society was not out of sight."[28]

The idea of implanting electrodes in human brains to control behavior comes closer to a potential application of "alien implant" technology. Jacques Vallée famously analyzes the Phenomenon from the point of view that it is a "control system."[29] What if his impression is literally true? What if it is a multidisciplinary control mechanism making use of everything from physically implanted control instruments to the use of UFO sightings as hypnotic devices designed to induce trance-like states, like those spinning disks seen on old science-fiction shows, or in the opening credits of *Twilight Zone*? What makes these admittedly fantastic proposals open to serious discussion is the fact—the amply documented fact—that the CIA (among other intelligence agencies in the United States and around the world) experimented with the ideas of implants, behavioral control, the manipulation of consciousness, paranormal research, wiping of memories, etc., as early as the 1950s; in other words, in parallel with the explosion of UFO sightings that had begun to take place after the end of World War II.

Is it possible that the "MJ" designation used in the (hoaxed?) Majestic-12 documents, which suggested the existence of something called MJ-12 for the study of extraterrestrial life forms, was a cryptonym only one step removed from the "MK" designation used for mind control at CIA?

Was MKULTRA an outgrowth of the "MJ" programs (or something like them) designed to study UFOs?

After all, much of the mind control/behavior control projects of the military and the CIA came out of the Operation Paperclip material,[30] particularly the data that was collected by aviation medicine pioneers including Dr. Hubertus Strughold and his team of Nazi scientists. The Nazis were associated with some of the earliest "flying saucer" prototypes, such as those by the Horten brothers. Is it such a leap to consider that there is a nexus between the consciousness research of the Nazi doctors on the one hand and the alternate energy and alternate propulsion research of the Nazi engineers on the other?

▼　　▼　　▼

The controversy over implants was given considerable media coverage when Dr. Roger Leir (1934–2014), a California podiatric surgeon, claimed he had removed these devices from several of his patients. However, none of these devices ever was examined by specialists outside of Leir's immediate circle, so no hard evidence was obtained to validate these persistent claims.

That does not mean that the experiencers are lying or being deliberately deceptive about their claims of alien-originated implants. The implant scenario may point to a different event entirely. It may be a metaphor for the psychic manipulation that takes place during an abduction: an invasion, or—as artist and psychic Ingo Swann described it—a "penetration."[31] The stories of invasive surgery and physical examinations generally precede or accompany accounts of alien implants. The documented instances of the CIA testing this type of device on human beings as part of the overall MKULTRA program is either an attempt by human beings to replicate something they had reason to believe was being done by the Visitors, or was itself mistaken for alien interference in human affairs.

Taken a step further, the implant meme may be related to the concept of "alien hybrids." This is a subset of the alien abduction scenario in which human women are impregnated by their alien abductors using artificial means to create a fetus that is part human, part alien; a hybrid of the two. This implies that human and alien biologies are related to such an extent that a hybrid is possible. This similarity in genetic and reproductive mechanisms would also explain how communication of any kind is possible between the two distinct species. It raises the question, however, of the origin and "homeland" of this alien species. Such a being would have had to develop and evolve on a planet or in an environment that was similar enough to Earth's that the resulting biological differences were minor: less than, say, between a human being and an ape. Presumably, genetic material of humans and "aliens" would have to be more than ninety-nine percent identical in order for this type of reproduction to be successful.

▼   ▼   ▼

It is worthwhile to point out here that Dr. Garry Nolan made quite a stir in 2018 when he published a report in a peer-reviewed journal[32] stating that the infamous "Atacama" skeleton, which was discovered in the Atacama Desert of northern Chile, was not a "hybrid" (as many UFO believers wanted it to be) but a deformed human child. This was based on deep genomic analysis of the specimen performed at Stanford University over a five-year period as described in the report, which was co-authored by fourteen other scientists. Much of the speculation about the Atacama skeleton being an alien-human hybrid or at the very least an alien itself was based on the rather unusual formation of the skull, which appeared quite elongated. To some this indicated a being of nonterrestrial origin. Many were so invested in the outcome that when Dr. Nolan and his colleagues published their findings, these observers were outraged; that can happen when an ideological position is at odds with scientific evidence. Many scientists would have loved to demonstrate that the Atacama skeleton was actually an alien, because it would have made their careers forever. Sadly, however, this was not the case.

Another key element of the "alien hybrid" scenario is the retrieval or removal of the fetus from the human mother at some point after conception but before birth. This means that the fetus undergoes a period of gestation in the human womb but is viable outside the womb at a very early stage. This theory is also on rather shaky scientific ground and would still require that the "human" and the "alien" in question be so alike genetically that it would be difficult to tell them apart.

Is it even possible, though—according to our understanding of science, genetics, and biology—that an alien race mated with humans at some point and produced hybrid offspring? Do we have to discount this possibility entirely?

It would be possible, of course, if the "aliens" in question were actually humans, as in the early CIA experiments in gene manipulation. We now know that *homo sapiens* mated with Neanderthals in remote prehistory, for instance, and produced offspring, descendants of which are among us today. The CIA program itself may have been inspired by stories of alien-human hybridization programs as recounted by experiencers. Would the CIA have reason to believe any of this at all, or would they have been intrigued by the possibilities offered by the accounts of the abductees? After all, Dr. Nolan and his colleagues spent five years studying the Atacama skeleton: there must have been the hope in the back of their minds that maybe—just maybe—they were looking at an extraterrestrial being or the offspring of an alien and a human.

▼   ▼   ▼

In either case—alien or human—we are presented with the uncomfortable realization that human beings can be regarded as machines that need tinkering, modification, and manipulation toward purely pragmatic ends: machines in the process of being reverse engineered. Indeed, futurist and inventor Ray Kurzweil suggested precisely this in an interview during the 2018 SXSW event in Austin, Texas:

> Kurzweil, who is currently director of engineering at Google, believes that clinical applications of biotechnology will profoundly transform health and medicine. One of the ways he said this will be accomplished is by improving one of the underlying components of our bodies: DNA.[33]

All of this is pointed out not to provide a rationale or theoretical support for the ideas of alien-human hybrids or alien implants, but to drill down on the concepts to understand what ideas might be lurking beneath the surface of this persistent theme. When matched against what was actually taking place—in the same country, at the same time (through the 1970s), and by agencies of the federal government and by their appointed and hired subcontractors including psychiatrists, engineers, and scientists, and perpetrated on its own citizens—the alien abduction scenarios do not seem very far-fetched. Instead, one could make the argument that the abductee phenomenon was a visceral manifestation of those same programs, a bleed-through of the secret projects into the dreams and fears of the general population.

It need not be interpreted as completely dire, however.

Whitley Strieber is one abductee who—for all the unpleasantness of his experiences—agrees that there is also something redeeming about it all, as well as something profoundly disturbing. These two emotional responses seem to be present simultaneously: to be spiritually elevated and, at the same time, emotionally devastated. This juxtaposition of emotions is evidence of the type of "high strangeness" that accompanies the Phenomenon, as if human beings are being made to experience a kind of psychological state for which there is no known precedent. It is this, more than anything else, that argues most strongly against the insistence of critics and professional skeptics that the UFO experience—especially the alien abduction experience—is nothing more than warmed-over sleep paralysis or a simple nightmare. One does not have to struggle to describe a nightmare, searching for the right words to limn the contours of the feeling of being chased by monsters or standing naked in front of one's peers or colleagues. The Phenomenon is not a space made for the type of pedestrian emotional responses with which we are all familiar in our daily lives. The Phenomenon creates its own space and makes its own rules. As the most famous observer of the Salem witch trials put it:

> Our dear neighbors are most really tormented, really murdered, and really acquainted with hidden things which are afterwards proved plainly to have been realities.
>
> —Cotton Mather, *Wonders of the Invisible World*, 1693

GODS, GENES, AND GENOCIDE

The people come here, Michael, to look for aliens, ghosts, and cults, and gateways to hell, secret military bases looking into other dimensions. I think, if there IS something, it is not none of these things–or perhaps all of them.

– *Resolution* (2012), Justin Benson
(writer, director), Tribeca Films

E ENDED *SEKRET MACHINES: GODS* WITH REFERences to Frankenstein and his Monster; to the golem; and to the tales of humans revolting against the Gods that created them. These ideas were introduced because they offer an important perspective on what is taking place today: an inexorable movement of science and technology that is obliterating barriers between what is human and what is artificial, a movement that tells us to embrace a New World in which it will become increasingly difficult to discern a human being from its simulacrum. And at the shadowy core of this movement—the Ghost in the Sekret Machine—is the Phenomenon, because what we are doing is establishing the parameters for our own evolution from Earth-dwelling creatures to space-faring instruments. We are creating the very robotic, androgynous beings that populate the nightmares of the exorcist and magician, the soul-shattering experiences of the alien abductees.

If consciousness is, as some scientists insist, inseparable from the brain, then the creation of artificial brains will result in the creation of consciousness with all its social and legal implications. It will challenge every conception we have of what it means to be alive, and aware, and human. It will relegate human beings to a corner of the created universe: vulnerable, primitive experiments and mere way stations along the path to mental and physical excellence. There has been much talk about claims of the reengineering of UFO machinery, but what if our robots, androids, and cyborgs are nothing less than reengineered *aliens*? What if it is not the craft we are reengineering, but the *pilots*?

This raises the inevitable question: Where do we draw the line between humans and machines? Between what is human and what is alien?

How much of who we are as humans is a construct, a device, a mechanism designed by another intelligence reigning on a dark throne in some mysterious palace in an interdimensional multiverse? Are aspects of ourselves mere components that have been cobbled together like so many spare parts from other units, other workshops in the cosmos? If so, what does that say about our consciousness, if it is truly nothing more than a secretion of our artificially generated brains?

The world is much more mysterious than we have been led to believe. Moreover, it seems that we have been aware of the secrets of our own origins for a very long time, perhaps on a subconscious level, and have not recognized them as such. Like our dreams, which—according to psychologists—may reveal hidden truths, our mythologies and our sciences contain very specific information from the "subconscious" events of our history. In fact, the patterns and symmetries we discover in nature suggest the involvement of an intelligence that sought not only to contribute directly to the creation of all life on our planet but, astonishingly, left a calling card behind.

One of our informants suggested that we look more closely at the role that genetics plays in the narratives associated with the Phenomenon. He suggested that we entertain the possibility that acts of genocide in human history had an ulterior function, if not an ulterior motive. In other words, there were some human genetic characteristics that have evolved over time that were less amenable to the much-rumored "alien hybrid" scenarios that contactees have reported with consistency over the past few decades. We were asked to consider that some ethnicities might be carriers of specific genes that were resistant to "alien" manipulation and that genocide was an alien-inspired or alien-directed program to weed out any potential problems: obstacles to alien control lurking in the human gene pool as it had evolved on our planet. Conversely, what if some of the genetic mutations that were "weeded out" of the human genome due to natural selection—i.e., in order to survive *on Earth*—may be necessary to enable us to live off-planet?

At first this sounds like warmed-over eugenics: the type of "race science" that justified the worst excesses of the Nazi regime. (Perhaps it is not for nothing that some contactees report the existence of so-called "Nordics"?) Looked at more closely, we can see that some races and ethnicities are carriers of genes specific to certain illnesses, such as sickle cell anemia among African-Americans and Tay-Sachs disease among Jews. What if there was another gene or genes that have no particular function or pose no discernible threat as far as we are concerned—perhaps hidden within our so-called "junk DNA"—but that would be considered dangerous to an alien race? In other words, what if the human race (or a certain ethnicity) was evolving "alien antibodies" that would render us invulnerable to an alien hybridization program, or whatever else the Others had in mind? What if we were developing intellectual or physical powers that would enable greater awareness of their presence, or provide us with some greater degree of protection against their intentions?

What if, knowing this, the Others have taken measures to ensure that these objectionable genes are not passed on or otherwise subject to even greater evolution or specialization? What if the polarization taking place between the races on Earth is part of this program and we are just playing along, oblivious to what is really happening?

In fact, what if the genetic code itself was an invention of these same Others?

Sounds outrageous, doesn't it? Pseudoscience and fringe conspiracy theory.

Humor us for a moment.

What follows is necessarily speculative. Although the data we present is genuine and fully supported by documents and by other primary source material as noted in the text, we are not suggesting that the conclusions we reach somehow reflect the "Truth," but rather serve as a platform for further study and research.

▼   ▼   ▼

The study of genetics and heredity is usually associated with the nineteenth-century scientist Gregor Mendel, a Catholic monk, who experimented with creating hybrid pea plants. As we saw in Book One, however, even Biblical authors were aware of the possibilities of creating hybrids between human beings and "angels." The Biblical account gives us an extra dimension to that scenario, however.

There was not only a new "being" created out of the mixed parentage of human and angel—in this case—but the certainty of conflict and struggle for survival. The episode of the "sons of God and daughters of men" in Genesis 6 is the immediate prologue to the story of Noah and his famous Ark: a means of preserving the genetic equivalent of every *desirable* plant and animal on the planet in order to repopulate the earth after the Deluge, a catastrophic event that would eliminate a large segment of the population, human and "hybrid" alike, from the world.

Genesis 6 was the first act of genocide mentioned in the Bible, and it was ordered and carried out by God.

▼   ▼   ▼

In the older, Babylonian stories of creation we are told that human beings were crafted as robots in a sense: artificial beings and programmed workers designed to make the lives of the gods easier. This only happened after a cosmic battle in which one set of gods defeated the other set and created humans from the blood of the slain. In another way of looking at it, human beings are a kind of cyborg, part divine and part something else, something not quite natural to the divine beings who created us, and therefore artificial. Humans are already a hybrid race from this perspective. The kind of "trans-human" species that is the subject of so much speculation in contemporary scientific and Ufological literature may be only a glyph of "trans-divine."

Therefore the idea of genetic heritage is a complex one, a spiritual history stained with the blood of battle and the enslavement of human beings. There may come a time in the very distant future when memories of the Holocaust become cast in this fashion: that there was a race of beings considered subhuman by one set of "gods" that were battling another set of "gods" and were enslaved and destined for destruction until the rescue of a small percentage of them. Think of Israel as a kind of Noah's Ark. If we can imagine this, then we can go back to the Biblical and Babylonian accounts—which are related—and reinterpret them as attempts to describe real events that happened in the distant past.

Since we began this project with the assumption that the Phenomenon is real, we are encouraged to interpret our history and our understanding of reality using this as the basic premise. There is a fundamental sentiment among many people that we are aliens ourselves: visitors on this planet, prisoners of this reality. We understand how bizarre human life seems to be, full of contradiction, violence, hardship, and a lifespan that is severely limited to at most a century, if lucky, and even then in a state of mental or physical disability for much of that time. Illness, war, famine, the hideous lives of many of the world's children . . . none of this makes any sense to us. We are conditioned to think in terms of purpose, and there seems to be no purpose in human suffering.

Are human beings an anomaly in the universe, or are we the natural products of a process that began as early as the Big Bang? In a sense everything that *is*, is because of the Big Bang; therefore, there can be no conceivable anomaly. All that exists—from a purely mechanistic point of view—exists because of a certain inevitability. We are writing these lines, and you are reading them, because of events that occurred billions of years ago and led inexorably to this point.

However, that is understating the case, isn't it?

For a very long time human beings were part of their environment. They lived in a symbiotic relationship with it. Gradually, though, humans began to see themselves as midwives to creation: sharing in the process of an evolving universe even as they were separate from it.

Zecharia Sitchin proposed that humans were created by the gods in order to mine gold on Earth. Without being that literal, we can say that metallurgy's origins *were* accompanied by a mystical view of the earth and of the evolution of metals. The historian of religions Mircea Eliade made a point of equating ancient metallurgical and mining practices with what would eventually become alchemy.[34] Alchemy itself is a "midwifing" process that unites human aspirations for spiritual perfection with the perceived perfectibility of metals, from lead to gold. Human beings, by working with metals and attempting to effect their transmutation, are essentially exteriorizing an internal process. As the interior and external processes become more and more aligned, changes are believed to take place in both the person and the metal.

However, when alchemy found itself discredited with the rise of modern chemistry and the other sciences, the idea of cooperation or collaboration with nature and creation was replaced with the hubris of control, domination, and manipulation. With that came the distancing of the essential human spirit from the process and from nature and creation altogether. Science became a way of interacting with the world that was purely intellectual, using mathematics as a medium. Suddenly, nature was at best a servant and at worst an enemy.

This idea had Biblical precedents. God famously had instructed Adam that he would have dominion over all the earth (notably not over the planets and the stars!). Before the expulsion from Paradise, Adam and Eve were simply guests who were told they could have anything they wanted except the fruit of a specific tree. After the expulsion, they were told their lives would be (to quote Thomas Hobbes) "nasty, brutish, and short." Giving birth especially would be painful.[35]

Gradually, over several thousand years, these ideas coalesced into a worldview—especially in the West but also, with the introduction of colonialism and its associated Abrahamic ideas, into African, Asian, and Latin American societies—that saw human beings separated from their environment in such a way that their presence on this planet seems almost contrived. The human condition—requiring food, water, shelter, and clothing, and which includes painful menstruation, pregnancy, and childbirth—seems alien to the planet. So much work to survive, not to mention reproduce and thrive. Most human beings live under harsh environmental, economic, and political conditions, and always have. Human relationships are fraught with drama and conflict, even those purporting to be love relationships between lovers, spouses, and family members. Thousands of volumes have been written to advise human beings how to live, how to relate to others, how to manage their lives. None of it seems to be very useful, at least not for very long. We take this state of affairs for granted; "that's life." Of all the creatures on the planet, we seem to be the most uncomfortable, if not the most miserable. It's a psychological condition known to philosophers as—appropriately enough—*alienation*.

There have been various attempts to rectify this condition, perhaps the most notable being the philosophy known as Buddhism. To a Buddhist, life is sorrow; the reason for this sorrow is attachment to the world, to ideas, to existence itself, even to the gods. All of these attachments must be neutralized in order to attain eternal bliss. In other words, one must leave the planet and even the universe itself.

There have been philosophies that face the problem head on, such as existentialism. There are religions that tell us, yes, the world is an awful place but it's the best of all possible worlds: a sobering thought, but at least this is only a temporary posting. We go somewhere better when we die.

(Thanks a lot.)

All of these ideas and religions and philosophies struggle with the same basic premise: that the world is not fit for human habitation. Not really. Somehow, we wound up on this planet, in this life, and while Earth seems tailor-made to support life (as we are told endlessly by scientists) it is not designed to support mental health or a peaceful existence. Our alienation from our surroundings is not the result of our contrary nature as human beings; it is due to something far more profound. While we can take joy in some things—a beautiful sunrise or sunset, the laughter of a child, the warmth of an embrace, music that touches our soul—these only serve to emphasize the ugliness of all the rest: the natural disaster, the drone strike, the crack baby, domestic violence, war, plague, famine, and drought. Life is beautiful. But it is also "nasty, brutish, and short."

Evolution is a fact. Granted, there are a few lacunae in the text; a few missing pieces, one or two assumptions, but nothing to get too exercised over. There are problems with understanding genetic mutations, for instance: we have been taught that evolution occurs as DNA becomes mutated and changes to accommodate the requirements for survival, but now we know that some (if not most) of these mutations have a negative impact, not a positive one, on evolutionary outcomes. In other words, these mutations cause us to lose capacities, not to increase them or expand them (except in certain, specific cases and under laboratory conditions). We also don't know quite how speciation occurs, or how quickly; just that it does. These observations aside, Darwin's theory of evolution is still dominant and still offers a comprehensive explanation that best accommodates the data.

Fundamentalist Christians and other religionists who have a problem with evolution see a disconnect between the age of the universe as determined by science and the age of the world as determined by Biblical analyses. They also insist that God directly created human beings (per Genesis) and that any suggestion that humans are descended from a long line of primates going back millions of years is an absurdity. They believe that the Bible is literally true and inerrant, and that discussions of creation, evolution, and of reality itself must refer to the core statements of that book.

There has been a tentative attempt to accommodate Biblical ideas with a kind of theory known as "intelligent design." Those who adhere to this theory claim that there was a Creator and that the world as we know it could not have come into being unless there was a Someone who designed it; that the structures of living things appear and behave with a certain symmetry, a mathematical elegance, which is evidence of an Intelligence behind it all. This is not a scientific theory in any commonly understood sense of the term, and indeed it was the brainchild of political and religious conservatives who wanted to promote the idea of God without actually coming out and referring to God as the "Designer." That there is a certain degree of congruence between this theory and that of directed panspermia (see below) usually is ignored, as the latter still does not require the presence of a Biblical deity.

This is what happens when ideology is permitted to drive the search for knowledge. The same phenomenon can be observed when political expediency dominates scientific research and development. In fact, it was a combination of politics and religious zealotry that closed down the military's remote viewing programs in the 1980s and was to blame for "pushback" from "high-ranking officials" against the US government's Advanced Aerospace Threat Identification Program (AATIP) because UFOs were seen as possibly *demonic*.[36]

Scientific research typically requires large infusions of capital. It is expensive to keep highly degreed scientists working for years on projects that may or may not bear fruit; it is costly to maintain state-of-the-art laboratories and manufacturing facilities, and to pay for the consumption of exotic raw materials, all in the hope that new technologies will be developed. This money comes from universities and large corporations—often defense contractors—and these funds originate from government budgets. The level of bureaucracy, red tape, and political intrigue involved in the allocation of these funds is mind-boggling. This means that some of the most "out of the box" thinking where science is concerned becomes toxic: one becomes involved with these theories at one's professional peril. Only those who have job security (in the form of tenure, huge government grants, or a Nobel Prize) can afford to advocate for new and challenging scientific theories. That is why a Stephen Hawking or a Michio Kaku can speculate on alien life, star drives, and the dangers inherent in the SETI project without fear of losing their status among their peers.

One of those high priests of the scientific community had done just that, and suggested that life on this planet did not spontaneously erupt at some point millions of years ago but that it literally was seeded here from some other place, some other planet, and moreover that it was done *deliberately*.

The scientist in question is Francis Crick (1916–2004), co-discoverer of the structure of the DNA molecule along with his colleagues James Watson (b. 1928) and Maurice Wilkins (1916–2004), who shared the Nobel Prize for the discovery, and Rosalind Franklin (1920–1958). Interestingly, Wilkins had worked on the Manhattan Project during World War II and was suspected by the FBI and MI6 of having been a Soviet spy (like Oppenheimer, Jack Parsons, Frank Malina, H. S. Tsien, and many others, as we will see in *Sekret Machines: War*). Crick, on the other hand, worked for the British Admiralty during the war and contributed to the war effort by designing a type of underwater mine that would defeat the German mine-sweeper, or *Sperrbrecher* as it was known (a technology that used magnetism to explode mines at a distance). Crick was able to design a mine that would be undetectable to the *Sperrbrecher* and would explode underneath it.[37]

After the war, Crick became involved in the study of genetics, mostly due to his friendship with Wilkins, who was both a physicist and a molecular biologist. Crick and Watson were not trained biochemists, but they were passionately interested amateurs who approached the problem of the genetic code by means of visualizing patterns and applying logical analysis.

The missing piece of the puzzle for Crick, even after they had "broken" the genetic code and discovered the double helix of the DNA molecule, was how a handful of amino acids could create life. The role of consciousness was something that obsessed him for decades after he won the Nobel Prize. Conversations with a colleague—the chemist and evolutionary biologist Leslie Orgel (1927–2007)—resulted in the two scientists proposing the idea that life as we know it came from elsewhere, by a process they called "directed panspermia."

Primer on the Structure of the DNA Molecule

Before we explore what that means, a short summary of what we know about the genetic code is in order. Briefly, the building blocks of all life consist of a number of chemical bases. These bases—only four in total—are known by their initials: adenine (A), thymine (T), guanine (G), and cytosine (C). Adenine can only bond with thymine, and guanine with cytosine, forming what are known as base pairs. These base pairs are attached to a sugar molecule and a phosphate molecule. Together—the base pair, sugar molecule, and phosphate molecule—are called *nucleotides*. They are arranged like the steps of a ladder in what has become familiar as the DNA helix (see Figure 1).

Figure 1. The double helix of the DNA molecule.

The base pairs produce amino acids, of which only twenty are considered "canonical" amino acids: that is, they are found in human DNA (deoxyribonucleic acid) and create proteins. Each amino acid is formed by a group of three base pairs, known as a *codon*. Thus, the amino acid glycine is formed by the three base pairs GGG (for guanine-guanine-guanine) or GGA (for guanine-guanine-adenine). The RNA (ribonucleic acid) molecule differs from the DNA molecule in that uracil (U) replaces thymine (T) in its composition, and it is usually found as a single strand rather than the double-stranded helix of DNA. RNA is often found as mRNA or "messenger RNA," as it relays genetic information to the cells. Tryptophan, the serotonin precursor, is represented by the RNA codon UGG for uracil-guanine-guanine. RNA is simpler in design than DNA, and some geneticists believe that the origins of RNA are earlier than DNA, based on the observation that complex systems derive from simple systems.

There are sixty-four possible combinations of the four base pairs in groups, or codons, of three each (4 x 4 x 4). This mathematical pattern is one that will be found in unexpected places, as we will see in a moment, when the reason for explaining all this becomes clear (see Figure 2).

Seond letter

First letter	U	C	A	G	Third letter
U	UUU Phe	UCU Ser	UAU Tyr	UGU Cys	U
	UUC Phe	UCC Ser	UAC Tyr	UGC Cys	C
	UUA Leu	UCA Ser	UAA Stop	UGA Stop	A
	UUG Leu	UCG Ser	UAG Stop	UGG Trp	G
C	CUU Leu	CCU Pro	CAU His	CGU Arg	U
	CUC Leu	CCC Pro	CAC His	CGC Arg	C
	CUA Leu	CCA Pro	CAA Gln	CGA Arg	A
	CUG Leu	CCG Pro	CAG Gln	CGG Arg	G
A	AUU Ile	ACU Thr	AAU Asn	AGU Ser	U
	AUC Ile	ACC Thr	AAC Asn	AGC Ser	C
	AUA Ile	ACA Thr	AAA Lys	AGA Arg	A
	AUG Met	ACG Thr	AAG Lys	AGG Arg	G
G	GUU Val	GCU Ala	GAU Asp	GGU Gly	U
	GUC Val	GCC Ala	GAC Asp	GGC Gly	C
	GUA Val	GCA Ala	GAA Glu	GGA Gly	A
	GUG Val	GCG Ala	GAG Glu	GGG Gly	G

Figure 2. The universal Genetic Code map showing the 64 possible combinations or codons, with U or uracil in RNA replacing T or thymine in DNA.

What occupied Crick and Orgel was identifying the point at which a collection of amino acids or their proteins becomes life. What was the transition from test tube chemicals to life, and from there to sentient life? Is consciousness nothing more than a wet bag of biochemical goo reaching some kind of critical mass? If consciousness is reached through evolution—gradually formed during stages from the single-celled amoeba to the brain of a Nobel Prize winner, or appearing suddenly as a side effect of the move from the oceans to dry land—then is consciousness itself evolving to something else? Is the next step in evolution the move from consciousness to a kind of super-consciousness, or to something altogether different and so far unimaginable, something paralleling our evolutionary move from the oceans to dry land as we now move from dry land to interstellar space?

Before they reached this level of speculation, however, Crick and Orgel were motivated by a purely scientific appraisal of the available evidence. They realized that a certain element—molybdenum—is necessary for the development of the genetic code and, in fact, is an essential element for life. Molybdenum is relatively rare on the surface of the earth (although it is more prevalent in the oceans). Crick and Orgel reasoned that the DNA molecule could only have developed in an environment where molybdenum was plentiful. Since molybdenum is a rare element on Earth, that meant somewhere off-planet.

The implication is obvious: a source from elsewhere generated the DNA molecule and transported it to Earth.

That did not mean that the selection of Earth was deliberate; our planet might have been one of thousands or millions of destinations for a wide scattering of DNA (or RNA) throughout the galaxy or the universe. From our perspective, that seems a terrible waste of resources. One would imagine that a more directed targeting of Earth—or perhaps of a handful of targets, all possessing the necessary basic requirements to enable DNA to replicate itself— would be more economical. Another possibility might be a comet hitting a planet that already had life, and pieces of that planet, carved off at impact, hurtling through space in the form of asteroids or chunks of rock, one of which hit Earth a few billion years ago.

It's possible that genetic material might have arrived on Earth in a microbial form via a meteorite. As it crashed into Earth—surviving a radiation-rich interstellar voyage and the perilous entry into Earth's atmosphere—it landed somewhere on the planet where there were sufficient nutrients, allowing the genetic material to spontaneously begin to replicate. All of this is what is meant by the term *panspermia*: that the seeds for generating life on this planet came from elsewhere.

These are only possibilities, and none of them have been proven, but Crick and Orgel tended to think in terms of a *directed* form of panspermia. By that it is meant that there was an intelligence behind the seeding of Earth with genetic material, and a deliberate selection of this planet by an organic precursor. The molybdenum requirement is one reason. The other is the structure of DNA itself.

If DNA was a purely random accumulation of amino acids, one would assume that there would be different forms of DNA in nature. However, the structure of DNA is the same regardless of the life form under discussion. In other words, whether we are talking about plants, animals, or human beings, the structure of their respective genetic codes remains the same (with only a few variations, mostly in mitochondria). This has led us to assume that it would remain the same wherever we looked in the universe. They are just different arrangements of the same basic "letters" of the code.

By way of analogy, we can think of the 0s and 1s, the "bits" of the binary code used in computer software. We can arrange the bits to form bytes (traditionally, a group of eight bits), use the bytes to write hundreds of different computer languages, then use those languages to create millions of different computer programs, which result in the software and apps that we use every day; but still, we can reduce all of that back to 0s and 1s. This is analogous to the great proliferation of life forms on this planet, all of which are reducible to the same basic DNA structure and composition. From a purely mathematical point of view this is quite stunning. The sheer economy of using this system to create all of life—from the smallest microbe to the largest mammal, every plant, every insect—is what seems to indicate the active involvement of an intelligence. That forces us, then, to ask a big question: From where did that intelligence derive this system, and does that intelligence share in the same genetic structure (is it identical to us?)

or was it created specifically for our planet, and for planets like ours? Questions like these prompt some people to think of a god or gods as the "prime cause" of life, but that solution reflects a lack of imagination. The transcendent God of the Abrahamic religions is still very far away from this scenario. We must address a lot of intermediary steps before we can begin even to think of a "god" in the traditional Western sense of the term. If there is or was an alien race that generated our genetic codes, then we have to go there first before making overarching statements about "God" and particularly God's direct intervention in the very localized Creation of the planet Earth.

In fact, ancient religions do have a "directed panspermia" myth. As pointed out by Robert Temple (the author of the bestseller *The Sirius Mystery*, first published in 1976) in an article[38] for *The International Journal of Astrobiology*, the idea of panspermia is very old and can be found in the Egyptian Pyramid Texts as well as in Sanskrit scriptures. The Egyptian god Atum, for instance, was said to have created the world from an act of cosmic masturbation, showering Earth with his seed. There are depictions of this act in several places in Egypt. Empty space was personified as the goddess Nu or Nut: she was the womb in which the spilled seed of Atum took root. In the Indian Vedas, we have similar stories (as Temple points out), and as one of our authors writes,[39] this type of episode is depicted in the Tantras as an act of seminal emission (or, in some texts, premature ejaculation) by the god Shiva.

Aside from these examples and the points we raised earlier in *Sekret Machines: Gods*—which refer to the textual evidence in the world's religions and ancient cultures alluding to contact—is there evidence of a more tangible nature to suggest that our genetic code has an "otherworldly" or off-planet origin? Were the ancients right in insisting that we are the creation of beings from elsewhere?

Crick and Orgel certainly argued for that explanation, starting with a presentation in 1971 and then extending to a co-authored paper in 1973.[40] While Orgel modified his position over the years, Crick remained a strong supporter of directed panspermia throughout his life.

More recently, two Russian scientists began arguing for the same position, but this time from a purely mathematical perspective. In order to appreciate their proposal we have to know that in addition to the twenty codons of the genetic code that we discussed above, there are three "stop" codons (UAG, UAA, UGA) and one "start" codon (usually AUG). Their function seems to be limited to providing a space between one group of nucleotides and the next, like a punctuation mark.

Or a zero.

▼　　▼　　▼

Mathematics is a language that represents quantifiable relationships in a symbolic way. It is the language of science, and often of philosophy as well. Some of the earliest known writings—those of the ancient Sumerians—are devoted to bookkeeping and accounting. Other Sumerian texts are concerned with astronomical phenomena, in lists with dates and approximate times the phenomena are observed. Numbers became a way of interpreting and organizing the objects seen and sensed in the world, and of drawing conclusions and making predictions about observed phenomena. It was easy to count the number of cows one had, or how many days to the next full moon. One could simply draw a line representing each cow or each day to be counted. That line was a symbol, a stand-in for the actual cow. This type of symbolic thinking and its representation gradually evolved into arithmetic and from there to the higher mathematics of geometry, algebra, trigonometry, and calculus. But before that could happen, mathematics needed another stand-in, another placeholder symbol, and that became zero.

The idea of "zero" is one of those accomplishments that helped to create the modern system of mathematics, and by extension the modern world. It helped differentiate 1 from 10, for instance. It could also represent the absence of a quantity, such as 10 minus 10. In modern mathematics, zero is considered an even number, divisible by two into +1 and -1 with no remainder (which may sound counterintuitive to some, but there you are). The Sanskrit term *sunya* means "desert," in other words, "emptiness," and it gave us—via Arabic *sifr* and Italian *zefiro*—the English word "zero" as well as "cipher." The ancient Mayans developed the idea of zero on their own (as far as we know), independently of the Indian usage, and can be dated to roughly the sixth century CE. The Indian concept of zero eventually made its way west, where it wound up in the writings of the famous Italian mathematician Fibonacci (Leonardo Bonacci, 1170–1250), who had grown up in North Africa and learned of its use in Arab and Indian mathematics there.

The reason for this brief history lesson on the number zero should become clear. We in the West did not know the concept of zero existed (much less create the actual numeral) until a thousand years ago. In Asia, it was known about five hundred years earlier than that. Before that time there were Chinese and Sumerian approximations of zero as a rather awkward placeholder, signifying an empty set, but that is pretty much the extent of it. "Zero" is something we humans came up with; we "discovered" zero or "invented" it when we needed it or something like it. It's artificial. It doesn't exist in nature.

Yet, from the point of view of two controversial Kazakh scientists, zero can be found employed quite deliberately in our genetic code; "deliberately," because within the RNA and DNA molecules the distribution of the start and stop codons—the placeholders that separate one set of nucleotides from another on the genetic chain—is *non-random*.

The entire theory of evolution is based on alterations taking place over millions of years and over many succeeding generations. Some of these "mutations" survive, due to the principle of natural selection, and others do not. We are the result of a long series of genetic mutations that have occurred over those millions of years. Each strand of

DNA in our cells is a history book containing every step we took, all the way back to the first humans who walked the earth, and earlier. Genetic testing companies such as Ancestry.com and 23andme.com specialize in taking a swab of your cheek or a minute amount of your saliva and returning to you an ethnic breakdown as well as, in some cases, probabilities of hair color, eye color, propensity to certain illnesses or addictions, etc. We watch television programs in which the wonders of DNA testing have resulted in the guilty being arrested and the innocent set free. Pregnant mothers can have genetic testing done to determine if their unborn child has certain diseases or disabilities. We understand that there is a certain order and predictability in genetics. It's a valuable tool for understanding our biological inheritance. For most of us, it's just there: an inherent part of our makeup as living beings. On a very elementary level we understand that when a man and a woman conceive, the resulting child will have a set of genes from each parent. That's basic secondary school biology. Sexual intercourse is about the extent of our conscious control of the process, and even then the possibility of conception from each act is relatively random. What we have *not* considered, however, is the possibility that the genetic code *itself* was created consciously and deliberately.

The Kazakh Configuration

In 2013, two scientists based in Kazakhstan—Vladimir I. shCherbak of al-Farabi Kazakh National University and Maxim A. Makukov of the Fesenkov Astrophysical Institute—published a paper in the journal *Icarus*[41] that claimed the genetic code contains the equivalent of "zero." This would appear to be highly improbable in a random evolutionary process, but it makes excellent sense if the code was created by an intelligent source.

The full thrust of their argument is too dense to go into in any great detail, particularly absent a good education in genetics and mathematics, and the original article is well worth studying. What we will do here is provide a much abbreviated version, with apologies to its authors.

They begin with an overview of the history of DNA research and the early attempts by several individuals to "decode" the genetic code using various mathematical approaches. While none of these attempts proved ultimately successful, they did manage to predict with a great degree of accuracy the symmetry of the code and marveled at its simplicity. With only sixty-four possible combinations of the four base pairs in groups of three creating the twenty amino acids and the start and stop codons, all of life on this planet—in its enormous variety and complexity—was created.

They specifically discuss the Gamow arrangement, which was developed in the mid-1950s. George Gamow (1904–1968) was a Russian theoretical physicist who studied at Göttingen (like so many of the most brilliant scientists of his time) and who was a friend of both Edward Teller and J. Robert Oppenheimer—the men largely credited with having produced the atomic bomb. Gamow defected to the west in 1933 from Russia and wound up as a professor at George Washington University in 1934. While there is no evidence that he worked on the Manhattan Project, he did consult with the US Navy during the war and represents yet another scientist who was part of the weird military-industrial cabal that we will be discussing throughout this project.

Gamow's understanding of the codons turned out to be in error, as he believed that each amino acid could be produced by its three bases in any order. It was discovered that this was not the case: the order of the bases is important. Since a change in the order of the base pairs would produce a different amino acid, this indicated that the "code" of the genetic code was "positional" much in the way that basic arithmetic is positional. The sequence 1-2-3 is not the same as 3-2-1 or 2-1-3, etc. The "position" of the integer is a critical component of the entire number. For nature to have come up with this system on its own seems counterintuitive, to say the least.

However, Gamow did see patterns and symmetries in the structure of the code, which was borne out by later analysis and research. The genetic code is quite specific: changing one "word" in the genetic sentence—whether by accidental mutation or deliberate manipulation—changes its meaning completely. Just as important, the "start" and "stop" codons are essential to the positional nature of the code, standing in as either punctuation or as the number "zero," depending on your preference. (Because the structure of the genetic code is more mathematical than textual, we may find it easier to refer to the "zero" analogy.)

Moreover, a mathematical approach to the structure of the amino acids shows that each acid is composed of a different number of nucleons or base pairs. Arginine, for instance, has 100 nucleons; lysine has 72; and so on. When these are added up according to the patterns devised by Gamow, the sums are strangely suggestive, offering up numbers as esoterically evocative as 666, 888, and 777.

Numbers may be said to rest at the very heart of the created universe. This was understood by Kabbalists as well as scientists. Mathematics is a language that we use to describe natural phenomena. We can observe everything from the famous Fibonacci series to fractals in seashells and snowflakes. But the symbols of numbers and the types of number systems are devised by human beings. The ancient Babylonians used a base-6 system; we use a base-10 system (a decimal system), but there are binary systems and sexagesimal systems as well. We use a system of nine numbers plus zero for our decimal system. If a zero appears after the number 9, for instance, it represents 90. If it appears before 9, it is simply 09, or just 9. That is an example of position. This is the same system that the Kazakh team discovered in the DNA molecule:

Note that all those distinctive notations of nucleon sums appear only in positional decimal system.[42] [*sic*]

They base their claim on both the arithmetic and what they call the "ideographical" nature of the genetic code, meaning the actual structure of the double helix of DNA and the single chain of the RNA molecule. (Indeed, the double helix structure of the DNA molecule has been pre-figured in religious and esoteric iconography from places and times as remote from each other—and from Watson and Crick—as medieval India and classical Greece.)

The fact that the DNA molecule contains start and stop codons that separate one nucleotide chain from another is only part of the theory. In other words, there is a kind of abstract construction in the genetic code that makes no sense if the code evolved naturally. Indeed, they address that possibility in their paper, demonstrating that it is statistically much more likely that the code was created deliberately rather than evolving as the result of a series of random mutations. It is as if whoever designed the code wanted to make sure that there would be no confusion as the genes evolved and grew ever bigger. It was a way of inserting organization into the genome, and that implies a deliberately structured process.

That is not to say that life itself was created this way. Life as we understand it could very well be the result of a random series of events, including natural selection, much the way scientists think of it today. However, according to *sh*Cherbak and Makukov, we should think of the first matter of living things, the primordial goo of chemical nutrients, as a biological medium into which—at some point in the remote past—the mysterious genetic code with its equally mysterious mathematical symmetry was inserted. This is the phenomenon they refer to as "Biological SETI" (after the more familiar Search for Extraterrestrial Intelligence, or SETI, which is done with conventional radio telescopes). In this way, the strands of DNA that make up all organic material on Earth can be considered either as mathematical equations or as sentences in a very long text. Like the letters of an alphabet, the codons are few in number but can code for very complex products: the equivalent of words, sentences, etc. However, the sentences would look like streams of consciousness without the judicious use of punctuation, and this is where the Kazakh Configuration (as we like to call it) comes in.

The key is the word "judicious." If the start and stop codons appeared at random, there would be no purpose to them. The result would be chaos. Imagine punctuation appearing at random (even in the middle of words) rather than at the end of sentences or phrases. Instead, the codons appear just where they need to appear in order to create the nucleotides necessary for life and for preserving the hereditary characteristics we associate with DNA. One cluster of base pairs may result in blue eyes; another, in a hereditary disease; another, in skin or hair color; another, in height. Considered genetically, rats and swine are only slightly different from humans. Hereditary characteristics are not the result of run-on genetic sentences but of discrete organization of the "text" at the genetic level. The start and stop codons keep the human genome "legible."

Further, there is an economy in the structure of the code that is breathtaking. Only twenty amino acids out of a total of sixty-four possible combinations of four chemical bases in groups of three . . . and all of life is created, from microbes to mastodons.

The Kazakh discoverers of this "Biological SETI" have been accused of supporting intelligent design or of being anti-Darwinist, but nothing could be further from the truth. They are scientists and mathematicians and find nothing in their work that would contradict either Darwin or the theory of evolution or natural selection. Instead, they focus on the moment when chemical constituents became carriers of life. They obviously believe that the *prima materia* of life preexisted the genetic code, and that once the code was in place, evolution proceeded normally.

How the code came to be present on Earth may be due—as the Kazakh scientists believe—to the directed panspermia described by Crick and Orgel. Some may have a difficult time accepting this possibility due to its strongly science-fiction flavor, but that is only the reflection of an anthropocentric perspective. When we insist on viewing all phenomena from a purely Earth-centered and human-centered position we neglect other possibilities. Even though it's now known that Earth revolves around the Sun, we still think the universe revolves around *us*.

If the mathematics of Kazakh visionaries seems too forbidding and their conclusions difficult to digest, we can cast a somewhat wider net. Remember what we said about the elegant simplicity of the DNA code, its sixty-four codons composed of all possibilities of four bases in groups of three.

What if we told you that this specific arrangement had been anticipated thousands of years earlier, on a continent far removed in both space and time from the field occupied by Crick and Watson? What if we claimed further that the arrangement had been understood as a code believed to rest at the heart of life itself?

And how would we explain that, once we had proved it?

THE CHINESE CONFIGURATION

The history of science is rich in the example of the fruitfulness of bringing two sets of techniques, two sets of ideas, developed in separate contexts for the pursuit of new truth, into touch with one another.

– J. Robert Oppenheimer[43]

MORE THAN THREE THOUSAND YEARS AGO, CHINESE philosophers had developed a strange system of encoded information that they could use to attain a deeper understanding of the universe. It was based on a binary mathematical system, with a solid line representing *yang* or the active, "masculine" principle of nature, and a broken line representing *yin* or the passive, "feminine" principle. The lines were grouped in eight sets of three, forming trigrams, from three solid lines to three broken lines and every possible combination between. The eight trigrams were then combined to form sixty-four hexagrams: "codons" of six lines each.

The system is known by the text that describes it, the *I Jing*, or "Classic of Changes." There are numerous arrangements of the sixty-four hexagrams that illustrate aspects of its mathematical symmetries, but they all can be applied to the structure of the genetic code.

For instance, in the Chinese numerological system the even numbers 6 and 8 are equated with the broken lines, and the odd numbers 7 and 9 are associated with the unbroken lines. These four numbers could be said to represent the four chemical bases of the genetic code (DNA): G—A—C—T. Since the Chinese binary system is composed of broken and unbroken lines, we can say—for the sake of argument and not insisting that the equivalence is meaningful—that G or guanine is the same as ▬▬ ▬▬. In this way A or adenine becomes ▬▬▬▬, C or cytosine becomes ▬▬ ▬▬, and T or thymine becomes ▬▬ ▬▬. Combinations of these four numbers in groups of three yield all sixty-four possible *I Jing* hexagrams, just as the four chemical bases in groups of three yield all sixty-four possible codons.

To illustrate this, revisit the chart of the genetic code (in Chapter 4) and simply replace the letters of the chemical bases with the broken or unbroken lines of the *I Jing* system. You will see that it fits perfectly. One can replace U with T—showing the DNA rather than RNA composition—and assign any of the "digrams" (the combination of two lines) to any of the chemical bases. It is possible to derive some "meaning" from these rather arbitrary assignments—finding a "stop" codon matching, for instance, a hexagram with the title of "Breaking Apart" or "After Completion"—but that is not our intention here. At this point, all we wish to point out is the perfect mathematical match between the *I Jing* system with its sixty-four hexagrams and the genetic code with its sixty-four codons. If the genetic code was seeded onto this planet at some remote point in the past, as Crick and Orgel posited, and if the code is mathematically elegant and evidence of deliberate intent, as put forth by *sh*Cherbak and Makukov, then perhaps we should not be surprised to see some consciousness of this reflected in art, spirituality, and culture. In fact, it is this very alignment of science and spirituality through the medium of mathematics that may be the key to understanding, finally, the true nature of the Phenomenon.

The *I Jing* is a divination system, which means it is used to predict future events. However, it is not the same as the standard sort of oracles with which we are all familiar. The Chinese classic is a philosophical text with profound statements about reality and about the relationships that exist between people and their environment. The *I Jing* is a study in dynamics: the hexagrams move from one to the other as broken lines become solid lines and solid lines become broken. The *I Jing* is about *process*, and the constant flow and movement of life. If we were to compose a poetic riff on the genetic code, what better metaphor could there be than the sixty-four hexagrams of this ancient Chinese text, which depict in ideographic form the sixty-four codons only recently discovered? Scientists resist finding meaning in life's events; to a scientist things just *are*. There is no meaning to be found, and none sought. The *I Jing* is a kind of response to that position, deriving meaning from the flow of one hexagram to the next.

And it is not the only system of its kind.

In Africa, among the Yoruba people of Nigeria in particular, there is another divination system, this one of uncertain antiquity. The similarity between the *Ifá* system and the *I Jing* is uncanny. This method uses the same idea of a binary notation system and is virtually identical to the *I Jing* system except that it is *even more complex*. Rather than a 4 x 4 x 4 = 64 hexagram system, *Ifá* is a 4 x 4 x 4 x 4 = 256 *odu* system. That is, there are a total of 256 different *odu* or binary forms that are derived mathematically in a consultation method similar to those of the Chinese practices. This is known as the *Table of Ifá*, and its practitioners must study for years before they can act as consultants to their people. That is because they must *memorize* the meanings, rituals, associations, and correspondences for *each* of the 256 separate figures produced by the Table. The *Table of Ifá* is a text, a scripture, that has never been written down.

Ifá is the repository of all the cultural, religious, and historical knowledge of its people. It is their clan's genetic code. Anyone who has been privileged to consult a genuine *Ifá* practitioner knows that peering into the Table is like disentangling one's own double helix, separating out the strands and reading them like entrails. Your own entrails.

The 256 figures are actually sixteen combinations of sixteen base figures (16 x 16 = 256) that are depicted as series of either two lines or one line, similar to the broken and unbroken lines of the *I Jing*. As such, it is another binary system of 0s and 1s.

The accompanying chart (Figure 3) shows the sixteen fundamental *odu* of the *Ifá* system with their Yoruban titles.

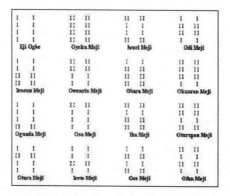

Figure 3. The sixteen fundamental odu of the Ifá system.

In the chart that shows the *odu*, the figure titled "Eji-Ogbe" symbolizes light, whereas "Oyeku Meji" symbolizes darkness. They may be considered analogous to the *yang* and *yin* concepts, respectively, of the *I Jing*. Eji-Ogbe is drawn with two columns of single lines, whereas Oyeku Meji is drawn with two columns of double lines. The other *odu* are composed of various combinations of single and double lines. This is the same concept that underlies the sixty-four hexagrams of the *I Jing*: combinations of single and double lines, or unbroken and broken lines, respectively.

An *odu* consists of eight separate "digits," as opposed to the six "digits" of the Chinese hexagrams. This adds another dimension of complexity to the system, a kind of fine-tuning. It is possible that this is a survival of an Arab or even a European divination system known as geomancy, which also consists of sixteen figures (albeit with different names; see Figure 4).

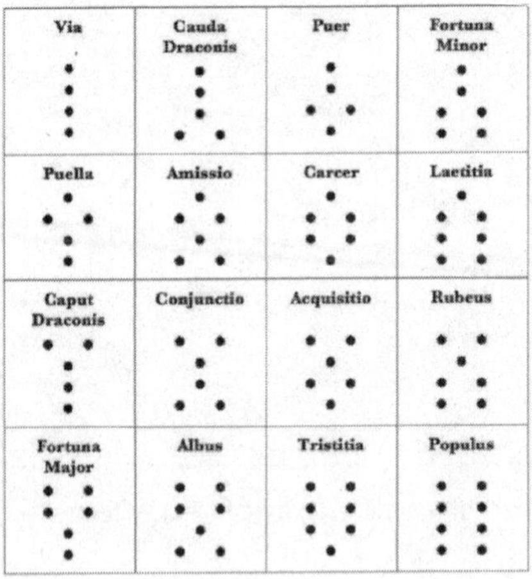

Figure 4.

You will notice that these figures are simpler than the Yoruban *odu* and may in fact represent a system older than the *Ifá* system, a kind of divinatory RNA rather than the full DNA of *Ifá*.[44]

There is a further level or organization that underlies the three systems—*I Jing*, *Ifá*, and geomancy—and that is the number 4. The geomantic system used by the Arabs and Europeans is a 4 x 4 = 16 system. The *I Jing* adds another level of complexity: 4 x 4 x 4 = 64. *Ifá* takes this one step further, to 4 x 4 x 4 x 4 = 256. Among Europeans and other cultural groups in North Africa and the Middle East, the number 4 had special significance as the number of the Platonic elements—Earth, Air, Fire, and Water—believed to be the building blocks of all existence (as the four chemical bases are the building blocks of the genetic code), but that may be an association that came after the fact. The Swiss psychiatrist C. G. Jung believed that the quaternary represented wholeness or completion, as is found, for example, in the designs of Asian mandalas with their emphasis on the four cardinal directions with corresponding gates, paths, gods, etc. Symbols such as the cross and the swastika are ways of expressing the idea of four as an ideograph of four directions, or of a sun or star spinning with four arms extended from its center.

But there is another system that accommodates all of these, a mathematical and linguistic marvel that was developed in the seventeenth century as a result of contact with nonterrestrial forces. Amazingly, it provides another mathematical element of the genetic equation; moreover, one that reinforces the "directed panspermia" theory. Like the Chinese, African, and Arab systems, it predates the discovery of the genetic code by centuries.

The Angelic Tablets

As will be discussed in *Sekret Machines: War*, the rocket scientist and cofounder of the Jet Propulsion Laboratory, Jack Parsons, was quite familiar with the work of John Dee and Edward Kelley, the two English magicians of the Elizabethan era who dabbled in ceremonial magic and mediumship. Parsons admitted that he employed their "angelic system" in his own occult workings. We have postponed discussing this system in any great detail until now, in the expectation that the foregoing material—particularly the genetic information as well as the brief overview of the Chinese and African systems—would enable the reader to approach this admittedly arcane subject with an open mind, or at least a curious one.

John Dee's reputation as a sorcerer may be a bit unfair, considering his expertise in navigation, astronomy, and mathematics and his role in introducing Euclid's *Elements* to an English-speaking audience (1570). However, it is his involvement with magic and spiritualism that is better known.

As we have been at pains to explain, it is necessary to reevaluate what we think of as occultism in light of what we have been learning concerning the Phenomenon. What was magic in an earlier time may be thought of as something rather different today: a kind of meta-science that incorporates elements of mathematics, physics, biology, and consciousness. Genetics may provide the key for understanding how these things are related.

Dee was a mathematician. So was Gottfried Leibniz (1646–1716) who, with Sir Isaac Newton (the alchemist and Biblical scholar who nevertheless is better known for his contribution to physics), invented the calculus. Leibniz, a philosopher and mathematician who made contributions to science that resulted in the creation of the first mechanical calculator, also discovered the binary system that is in use today in computer technology and was the first to recognize it as the basis for the *I Jing* (Figure 5).

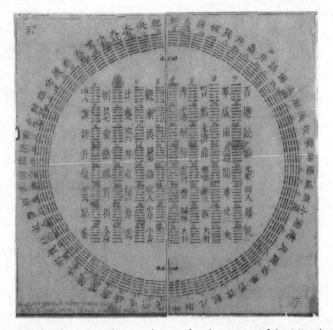

Figure 5. This document, showing the sixty-four hexagrams of the *I Jing*, both in the square matrix format and as a circle, was owned by Leibniz himself.

Dee, however, was most likely unaware of the binary system. His discovery was much more complex and has resisted attempts by believers and skeptics alike to deconstruct it or explain it away as the deranged symptom of a superstitious age. While there are a lot of moving parts to the "Angelic Tablets," we will focus on its mathematical construction and in particular the way it was employed by people like Jack Parsons, who studied it in its more contemporary interpretation.

The Angelic Tablet (see Figure 6) consists of a large square divided into a grid of 675 cells (25 x 27 = 675). The Tablet may be subdivided into four smaller tablets of 156 cells (12 x 13 = 156), each of which contains a letter. The smaller tablets are associated with the four Platonic elements. Thus there is an Earth Tablet, an Air Tablet, and so forth. This is another iteration of the quaternary, or number 4.

There are fifty-one cells left over, forming a grand cross between the four smaller tablets. However, only forty of these remaining cells contain letters, and these repeat. By eliminating repeated letters in those forty, we are left with a small tablet of only twenty cells (4 x 5 = 20). This is called the "Tablet of Union."

Of the smaller tablets of 156 cells each, some form crosses of ten cells each. There are four of these crosses in each smaller tablet, giving a total of forty cells. According to the system of the Golden Dawn (a British occult society of the late nineteenth century that based its system of magic on the work of Dee and Kelley), the crosses are referred to as "Calvary Crosses."

Across the horizontal arm of each Calvary Cross there are an additional four cells, for a total of sixteen cells per tablet.

What are left are sixty-four cells, each with a letter, and each with a particular "elemental" association. For example, in the Air Tablet you will have an Air "column" and an Air "row" or "rank." The cell that occupies the Air rank in the Air column of the Air tablet is purely "air." Thus, you can overlay the arrangement of the hexagrams of the *I Jing* directly over the Angelic Tablet using the same system of elemental associations.

This means, of course, that you can also overlay the sixty-four codons of the genetic code over the Angelic Tablet.

This may seem like a coincidence or maybe wishful thinking, but the mathematical symmetry is there.

However, we have four of those tablets, which gives us a total of 256 cells (4 x 64 = 256). This means you can overlay the 256 *odu* of the *Table of Ifá* on the Angelic Tablet, using the same binary and elemental system we have been discussing. The systems all fit each other nicely and indeed are based on some similar assumptions, including the primacy of the number 4 as the organizing principle of all of these systems.

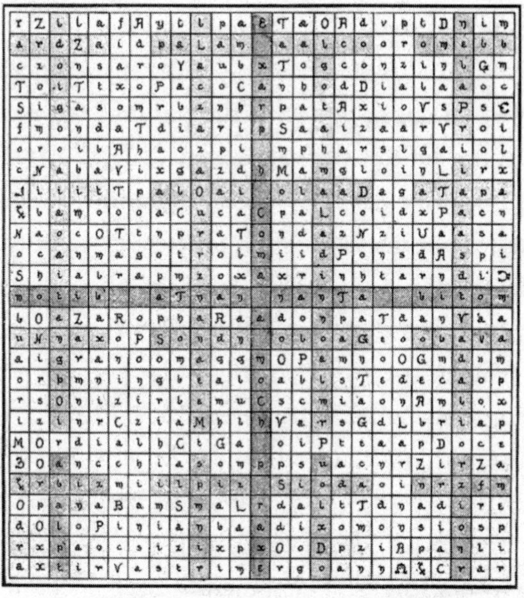

Figure 6. This is one version of the original Dee tablet. The squares that make up the Tablet of Union are shaded: one horizontal arm and one vertical arm.

The Angelic Tablet devised by Dee and *used for communicating with extraterrestrial beings* and expanded upon by the Golden Dawn, however, contains one more important clue: the Tablet of Union.

As we mentioned, the Tablet of Union consists of twenty cells. These twenty cells are taken from the Grand Cross that unites all of the four tablets that make up the greater Angelic Tablet. We propose that these twenty cells are an unconscious perception of the twenty amino acids that are produced by the sixty-four codons.

Yes, this is an outrageous (and unproven) claim. But look at the numbers and the patterns they produce. It is no less outrageous than seeing an intelligence at work in the creation of the genetic code itself. The number games that were played by Gamow and, later, by *sh*Cherbak and Makukov are consistent with the patterns we see emerging from the ancient divination systems and from the seventeenth-century Angelic Tablets of John Dee as well as their later interpretation by the nineteenth-century British secret society the Golden Dawn. It is as if knowledge of the genetic code was slowly emerging all over the planet—in China, Africa, the Middle East, Europe—and becoming increasingly articulated and defined until finally the code arose from the work of Watson, Crick, et al., in the twentieth century. This is a phenomenon that defies rational explanation unless it is linked with directed panspermia and thereby with the Phenomenon itself, demonstrating a link between directed panspermia, an alien intelligence, and consciousness.

The double helix structure of DNA—and the single twisting strand of RNA—was not known to science until Watson and Crick's discovery in the mid-twentieth century. Yet the ideographic elements of both were known in esoteric circles for thousands of years. The familiar symbol of the caduceus—two serpents entwined around a central axis—was the emblem of Hermes, the Greek god we have come to know from the discussion of Hermetism in Book One. Hermes was the messenger of the gods who became associated with the deep mysteries of initiation and transformation. The caduceus sports another feature that may allude to its divine (or, at least, nonterrestrial) origins: wings. The serpents of the caduceus are winged serpents.

This same ideograph is familiar to students of yoga and Tantra. It is used to represent the twin serpents—the Iḍā and the Piṅgala—that are entwined around the central column (analogous to the spinal column) of the human body, the Suṣumnā. The Iḍā and Piṅgala are components of a binary energy system within the body that are often identified with the *yin* and *yang* (respectively) of Chinese medicine and mysticism. The deliberate balancing of these two channels through yoga and other practices contributes to an increasing integration of the body's many internal systems (including the autonomic nervous system, the hormones, etc.), resulting in complete biological and psychological unity: a state that may be permanent or temporary, depending on the practitioner.

Some of the earliest examples of the caduceus come to us from ancient Sumer. The famous icon of Ningishzida, the Sumerian underworld god, shows the god as two serpents entwined around a central axis, as seen in Figure 7.

Figure 1. The Serpent Lord

Figure 7. This depiction of the Serpent Lord is four thousand years old.

Another intriguing representation of the twin serpent motif was discovered in China, in Xinjiang Province (in the far western part of the country, bordering on Central Asia). It is of uncertain date, but depicts Nu Wa, the Chinese goddess who created humanity, and her brother Fu Xi, who assisted in the creation of human beings and was considered the "first human." These are the gods credited with creating the human race, and the painting depicts them as two serpents intertwined and surrounded by symbols representing the constellations. To make matters even more interesting, Nu Wa and Fu Xi are holding what could be interpreted as Masonic instruments, as seen in Figure 8.

Figure 8. The origin of humanity, depicted as a double helix coming from the stars.

While rare, a single helix structure may be found in some natural objects—such as certain seashells or some climbing plants—but a double helix seems to occur naturally *only* as the structure of the DNA molecule. Like the "zero" in the Kazakh Configuration, it does not appear in nature except in the DNA molecule. Yet this design—the ideographic aspect of DNA—was used as a symbol of secret wisdom in cultures as far apart as Greece, China, Sumer, and India.

The symbol for RNA was prefigured in the single helix emblems of the Staff of Asclepius, the symbol of medicine (Figure 9).

Figure 9. The Staff of Asclepius.

Asclepius was a Greek god of medicine who was actually a hybrid himself: the son of the god Apollo and a human woman. The woman was killed because she was unfaithful to Apollo, and her unborn child—Asclepius—was cut from her womb.

▼   ▼   ▼

One may think that all this is a pastime of intellectual elites, a Glass Bead Game devised for entertainment or maybe a rarified means of education. However, the ubiquitous nature of this system is present all over the world in a much-reduced and much-simplified form: the common chessboard (see Figure 10).

Figure 10. The common chessboard is also the perfect matrix for the genetic code, the "magic square" of Mercury (the messenger god represented by the caduceus, symbol of RNA/DNA), the 64 hexagrams of the *I Jing*, and a host of other relevant associations.

Each chessboard consists of sixty-four squares, or cells. Thirty-two of these squares or cells are black; the remaining thirty-two are white or red. They are a battlefield on which is played a cosmic struggle between kings. Children learn how to play checkers or chess on this board. At the same time one can overlay the Chinese hexagrams or the genetic codons directly onto the sixty-four squares, to make the game a little more interesting.

The term "checkmate"—used to represent the end of the game of chess in which one side is defeated—comes from the Persian *shah mat*, or "the king is dead." How did it happen that the number sixty-four—so pregnant with meaning and positive associations concerning the origin of life on Earth—came to represent a battlefield on which two opposing forces confront each other to the death?

King's Gambit

The subtitle of the third volume in this trilogy is *War*, and for good reason. It is about authority, kingship, bloodlines, and an ancient and eternal conflict. It is a story replete with dark forces, super weapons, colonization, slavery, abduction, and more. There are gods and humans, and maybe a composite or hybrid race. There are secret tunnels, flying machines, interdimensionality, and a multiverse.

Conflict is at the very heart of the human experience, and we have enough sensitivity to realize that something is wrong with this picture. As we have said before, at the very outset of this project, human beings are reliving and revisiting an ancient trauma, over and over again. Religion is one of the ways in which we relive that initial contact.

War is another.

Before we explore all the ramifications of war and conflict with reference to the Phenomenon in the next volume, let us look at the biology of conflict and how, as it turns out, we learn it from an early age. A very early age. Think prenatal.

Most people are aware of the dangers of pregnancy to the mother. It doesn't seem "natural" that this most natural of processes should be potentially lethal, as if life and death were bundled in a single package. Other creatures, even other mammals, do not risk obliteration because of reproduction, but human beings do.

The human fetus is a greedy little creature. It competes with its mother for nutrients, oxygen, and blood. It latches its placenta onto the mother's endometrium which is rich in blood and nutrients; it then immediately begins to control the mother's blood supply and increase her blood sugar and blood pressure. It also uses the mother's body to remove waste material from the embryo. It engages in a life-and-death struggle with the very entity that conceived it. It manipulates its mother in incredible ways, including playing with the hormonal system and even sending fetal cells into the mother's own blood supply. It is a wonder that she even survives the process at all.

That may seem like overstating the case, but every year it is estimated that over 500,000 women in the world die in childbirth or during pregnancy, and that millions more are affected with injuries and disabilities due to pregnancy. Fully fifteen percent of women suffer complications from pregnancy that are life-threatening.

Other mammals do not have this problem. Pregnant females are able to carry on daily activities and even evade predators, give birth, and then get up and go about their business shortly thereafter. Human females, on the other hand, are an exception, and we don't know why.

What we do know is that the human fetus does what it can to control the mother. It is actually quite aggressive. Its placenta attaches itself to the mother's endometrium and begins to commandeer her blood supply. The human mother's endometrium is quite tough when matched against those of other mammals; it has to be, for the placenta is aggressive and will do all that it can to absorb as much of its mother's blood as possible.

On one hand, the endometrium needs to restrict the intake of the fetus in order to protect the mother; on the other, the fetus needs to absorb as much blood and as many nutrients as possible in order to survive and thrive. It's a struggle between two living creatures over the available food supply, and it's this struggle that leads to complications during pregnancy, such as eclampsia (high blood pressure in the mother due to the excessive demands made on her blood by the fetus).

The human menstrual cycle is one result of this unique biological condition. While menstruation is not unique to humans, the toll it takes on the human female is considerably heavier. The functional layer of the endometrium—which is built up during every menstrual cycle—is flushed out during menstruation. This serves several functions, notably to remove any nonviable embryos as well as the unfertilized ova from the mother's body. A quantity of blood is also flushed, while some of it is reabsorbed into the mother's body. Other mammals experience menstruation, including primates as well as a type of bat; in the case of humans, however, the menstrual cycle carries with it a great variety of disorders from headaches and cramps to more severe bleeding, mood swings, etc. Thus a menstruating woman is clearly at a disadvantage when it comes to predators, so it would seem that evolution would not favor such a condition. Yet here we are.

But what does all of this have to do with chess, aliens, and the genetic code?

The fetus and the mother are opponents in a game of chess. The chessboard is the interface between the placenta and the endometrium, a region where blood, hormones, and gases are exchanged, along with a variety of nutrients. Each side wants to survive at all costs. And the most powerful piece is the queen.

The optimal outcome in this game is the draw, in which neither side "wins." A child is born with no harm to the mother. The child is itself a chessboard with two sets of squares, or cells, and two opposing armies: the mother's DNA and the father's.

There is the phenomenon of genetic imprinting, in which genes from the father cause certain functions in the fetus's DNA to turn on or off. From an evolutionary point of view, the father's interests are not the same as the mother's. The father wants his child to come to term, regardless of the effect on the mother's health. The mother, after all, can become pregnant by another man, an outcome in which the first man has no interest. This competitive aspect is emphasized to the extent that the health of the fetus and its ability to thrive is believed to be linked to the degree of closeness between the mother and the father. In short-term relationships—one-night stands, for example—the fetus is believed to be weaker in some respects due to the lack of certain genetic traits that are otherwise present in a fetus conceived as a result of a long-term relationship. This sounds impossibly cheesy, but in fact this is the prevailing belief among biologists today.

The idea of the placenta/endometrium interface as a matrix is not uncommon. Even the term "matrix" derives from the same root as "mother" and "matter": the Latin word for "womb." A chessboard is a matrix of 8 x 8 cells, just as the various arrangements of the *I Jing* hexagrams appear in 8 x 8 cells. The genetic code is based on a matrix of 4 x 4 x 4 bases. These are all "wombs" of the material world, a world considered both as physical matter and as the processes that comprise it. The matrix is a pattern on which the world we experience is formed. It is not enough for us to know about physical laws as such, but also the *patterns* that shape the physical world. We exteriorize this understanding through game play, such as chess and checkers or their ancient antecedents, such as the Indian game of *chaturanga*, which also is played on an 8 x 8 matrix. As in the proposal by *sh*Cherbak and Makukov, the math behind this is both arithmetical and geometrical; in other words, there is not only a computational aspect to the code but also an ideographic (pictorial, symbolic, pattern) aspect to it. The 8 x 8 matrix we use to determine all possible combinations of the four bases in groups of three in *two*-dimensional space becomes the double helix pattern in *three*-dimensional space.

Yet for all of this basis in mathematics and physical laws, it is not only "matter" (matrix) that is formed during pregnancy; a new consciousness is also introduced into that matrix. Consciousness may be linked to the new genetic formula represented by the sharing of the maternal and paternal chromosomes, as an element not only of chemical and biological material but also of its position in space and time—external environmental factors whose importance cannot be ruled out. Every chess game is different.

▼   ▼   ▼

We are taught that most of what we experience in Nature is the result of evolutionary processes that have been taking place over millions of years. Every plant, animal, insect, microbe, and human being share a common genetic heritage. The concept of natural selection indicates further that those whose gene pool survived those millions of years—the person reading this text now, for instance—behave in a way that is consistent with evolutionary requirements.

The driver behind all of this is DNA. DNA wants to survive and to reproduce. In a sense, human beings are nothing but a transportation and reproductive medium for DNA. As mentioned earlier, Richard Dawkins has called DNA "the selfish gene." DNA is also theoretically immortal. As long as it keeps reproducing, it will never die. We contain all the DNA of our ancestors; we are walking libraries of genetic history. Our biological structure—from skeleton to organs to brain and nervous system, everything—is designed for one purpose only: to pass the genetic code down to succeeding generations.

There is genetic material in your cells that is millions of years old, which is as good an indicator of immortality as we can expect.

One theory has it that, as individuals, we are not expected to survive; there is no profit in our individual survival as far as DNA is concerned. As long as we have reproduced, there is no advantage in keeping us around. We are expected to live only as long as we are able to keep reproducing; after that, we have to die in order to free up resources for the next generation. Immortality of the individual—from the perspective of the genetic code—is counterproductive. Diversity of the gene pool is the goal, and if an individual were to live for a thousand years, his or her genetic material would be redundant. An immortal human being—from the point of view of DNA—would be a waste of space.

It's not that senescence and death are programmed into the code. Death can come in many forms, from violence and war to natural disasters. While we may be able to manipulate our genes to the extent that disease becomes a thing of the past, there is always the threat of attack by other humans, or the occurrence of a simple accident. The code has made allowances for that, as well. Humans, by the very nature of their environment, do not last long. Thus there is no point in building a human that could last hundreds of years under optimal circumstances if those circumstances are impossible to anticipate or produce. It's not that death is programmed into the genes; it's just that immortality is not. Eventually, the genes in the body stop replicating and repairing life's damage to the organs because the body is most likely on its way out.

Therefore, immortality for human beings has to be the result of a deliberate choice. We can't count on our bodies to do it for us. The genes have no illusions about living forever (or even just for a very long, Biblical-style life expectancy, like Methuselah), but we, as individuals, have a different set of values. Somehow our consciousness is at odds with our genetic inheritance. When it comes to life and death it seems we don't agree, and that could be the result of consciousness coming rather late to the party.

In the reported cases of alien abductions, it seems the aliens bypass our consciousness as if they are not impressed by it or they consider it unimportant. They go straight to the organism, the body itself. Human consciousness is something they want out of the way. It seems to interfere with their plans.

Much has been made of the purported alien interest in human (and animal) reproductive organs. It seems odd that supposedly advanced creatures who can travel back and forth to Earth from their home planet in the twinkling of an eye would be so backward technologically that they would practice crude forms of surgery on abducted human beings. What may be taking place, however, is something more profound.

If, as we suggest, our genetic code came here from elsewhere, it stands to reason that it developed along evolutionary lines that are quite different from those on its planet of origin. It may even be that RNA was the seed used to populate life on Earth, and that DNA is itself a kind of mutation. Today, RNA seems to serve functions that only are in service of the DNA molecule, but what if that represents a development that confuses or puzzles the Others?

Further, what if the Others' own genetic evolution has reached some kind of end stage? What if, due to inbreeding over millions of years, their DNA has become nonviable? In a sense, we are in danger of the very same thing, although we may not see the results for another million years or so. We normally think of inbreeding as something that takes place within a close family or among people who are related, or in small ethnic communities that do not procreate outside their circle. Earth, though, also is just such a community, despite the fact that it is quite large and contains billions of people, most of whom are not closely related. One can imagine that eventually—with climate change, wars, famine, drought, epidemics, and the like—Earth's population will become smaller. Those people who are left will be carrying genes that are similar enough to the other survivors that the speed of negative mutations will increase. The only remedy would be to incorporate genetic material from other sources, off-planet. Although we are on the verge of manipulating the genetic code as easily as computer code, there is always the law of unintended consequences to consider. One serious mistake and the planet could become prey to a rabid virus or a bacterium for which there is no cure. (Indeed, most of our DNA is actually bacterial in origin.) As it stands, we are already in danger of succumbing to super-bugs that are resistant to any of our antibiotics.

Let's say, for the sake of argument, that something very similar to this has happened already to another species on another planet, or perhaps to a species somewhere hidden on our own planet (known in the literature as "ultraterrestrials"). They know that our genetic material comes from the same place as theirs; we are related. Of course evolution has caused enormous changes in our DNA over the millions of years we have been here, but the basic code is the same and replicates the same proteins, only perhaps in different combinations.

Let's go a step further and say that this other species wants to "reset" their DNA because they've found themselves on an evolutionary dead end. They are in danger of becoming extinct on their planet the way Neanderthals and Denisovans have on ours. At the same time, they don't want to lose any of the characteristics they value, characteristics that they have and we don't. They also don't want any of our vulnerability to certain diseases (perhaps even diseases we don't know exist or are coming our way). They don't want any of the traits that they consider negative in terms of their environmental requirements. Let's say they have tinkered with their own genetic material, excising so many genes they felt were negative mutations that now their ability to evolve and survive has been drastically reduced, perhaps in relation to environmental factors they now experience but which didn't exist before they started tinkering.

What they need to do is obtain genetic material from another source, alter it to satisfy their requirements, and experiment to see if—taken to term—a viable offspring is produced. They can't do this on their home planet for some reason—perhaps their native environment has become too unstable, or they are lacking in appropriate nutrients—so they have turned to our planet. Obviously, our genetic material is robust enough to have survived millions of years. But their environment and ours are different enough, and there are so many unpredictable variables, that the only way to discover whether their approach will work is to "farm" embryos that are part alien and part human.

As this book was being prepared for publication an important news story came out of China. Chinese scientists have successfully edited the genes of nonviable embryos in order to neutralize the gene considered a marker for the HIV virus. The previous year, another Chinese team had used the same technique to modify the gene responsible for blood disease. This technique—known as CRISPR (*clusters of regularly interspaced short palindromic repeats*)—is a revolutionary editing tool for genetic material that is used to remove sections of a gene and replace them with other genes, a lot like editing a text. If *we* can do this now, the scenario of alien abduction and the creation of alien-human hybrids becomes more plausible.

CRISPR technology is not difficult to come by. One easily can imagine an unethical operation in a country ruled by a dictatorship authorizing and even encouraging experimentation on viable embryos, perhaps abducting women to serve as unwilling participants in these experiments. It is easy to see that had this technology been accessible to the Nazi eugenics scientists, they would have used it.

If an extraterrestrial or ultraterrestrial race lives among us, they might feel compelled to do the same without our knowledge or agreement. Their survival would be at stake. They may look upon us the way we view lab rats and guinea pigs. We have no way of knowing. All we have are statements from decent, well-meaning, perfectly sane individuals who claim that this indeed has happened to them.

On Earth, between nations, such action would be considered a *causus belli*: a reason for war.

In this case we have an enemy that is largely invisible to us. There is nowhere to point a weapon or wage a battle. The enemy possibly lives among us, disguised in some way. Or they come from a place unknown to us, far from the reach of our technology, and can swoop down and interfere with us at their will.

In a sense, they are like terrorists.

We would say they are accidental terrorists. It is not clear that terror is a tactic they employ, but merely a side effect of their actions on the rest of us. Appearing and reappearing. Abducting and releasing. Mixed signals and hybrid babies. Now you see them, now you don't. A Scarlet Pimpernel of reality.

When Columbus crash-landed on American soil, he was with his crew on ships that the Native Americans had never seen, dressed in clothing that was strange, speaking an unknown language, and possessing technology that seemed like magic. He had even witnessed a UFO on the night before his landing in the Bahamas—on October 11, 1492—as if a delegation from a hidden base on Antarctica or a fortress on Sirius had decided to give their somewhat crooked blessing to his endeavor. We don't know what those Native Americans really thought of Columbus and his ships; within years they were wiped out, due to foreign illnesses against which they had no natural immunity, and to the mortal cruelty of Spanish slavery.

A cautionary tale, perhaps.

▼　　▼　　▼

There actually is another type of cell that is immortal, as long as it has a host. Cancer is theoretically immortal. Cancer cells will reproduce constantly as long as their host body remains alive. Cancer cells have obviously not learned how to balance their reproduction with maintaining the survival of their host, however. Should cancer learn how to do this, one wonders if a cancerous tumor would eventually develop a kind of consciousness, and if it would be completely separate from the host's consciousness.

Both the DNA molecule and the cancer cell are immortal and desire to reproduce as much as possible; the difference is that cancer cells cause the death of their host, which means their own death. Cancer cells, while theoretically immortal, are also suicidal.

The DNA molecule also has very little regard for its host. As we have indicated, the immortality of the host is not its goal. Human beings stay alive only long enough to reproduce as much as possible and to ensure the viability of their offspring. We have created strategies for prolonging life beyond that point, but in the process have encountered new problems: the appearance of illnesses in old age that did not exist in youth. As the body ages, its internal support systems begin to break down. Psychological as well as physical disorders appear. We become less ambulatory; our world is reduced to our home, then one room, then finally our bed, which we know will become our bier. Life, difficult at the best of times, becomes increasingly unbearable.

At a certain point it would seem that our genes want us dead.

Consciousness rebels against this state of affairs. Consciousness is no friend of the genetic imperative. Consciousness *does* desire immortality, even convinces itself that immortality exists and is possible and comes up with all sorts of strategies to attain it. Every culture is aware of immortality—in their myths, legends, spiritual practices— even though there is no evidence of its existence. We posit the realm of immortal beings in heaven or on the summit of some Mount Olympus or Mount Kailash. Christianity is based on the idea of resurrection: of coming back from the dead. This was the "super power" of Asclepius, the Greek god of medicine. He was able to bring human beings back from the dead.

Shamans go through a death-and-rebirth ritual, coming back to their community with their own set of superpowers. It's a ritual that involves traveling to the stars. Ancient

Egypt is the home of the resurrected god, Osiris, and the process of mummification, which is studded with references to the soul existing in a realm after death, traveling to the stars. All of these motifs involving death, resurrection, and interstellar travel that we discussed in Book One now come back into the discussion, but from a different perspective. We have shown that scientists—not speculative historians—entertain the distinct possibility that our genetic code was designed by an intelligence and that the code came from the stars and was seeded here deliberately. The encoded information in our cells has been trying to get our attention for thousands of years through divination systems in China, Africa, Europe, and elsewhere that are based on the same mathematical pattern.

As we evolved to a certain level of sophistication according to the encoded program in our cells, contact was made with us. This contact was either from those who had created the genetic code and seeded it onto this planet or from another race of beings entirely. In any case, and for the sake of argument based on textual and other evidence, we suggest that this contact was made by a race that was related to us genetically. That would explain the similarities in body structure and the fact that communication of a sort took place between us. They were different enough in appearance, however, that our representations of them show that they were either much bigger or much smaller than ourselves; that they had characteristics that we associated with various animals; that they had abilities we understood to be paranormal. And they had mastered flight.

Right or wrong, we characterized this contact in supernatural terms and identified these beings as the "gods." We identified their place of origin as the "stars." And we knew, somehow, that they had "created" us.

All things being equal, that should have indicated that we—the human race—was well on its way to creating a heaven on Earth. We had the exemplars of our evolutionary path in front of us: supernatural beings with tremendous powers who could enter and leave our world at will. It was only a matter of time.

But then something went wrong.

▼　　▼　　▼

The term "ghost in the machine" is used to describe the mind-body problem. Philosophers with a materialist bent propose that the mind cannot exist outside of the body; that the body can exist in a vegetative state without any discernible brain waves, but that the death of the body means the death of the mind, as well. The body is the machine; what we call the mind is really just a ghost in the machine.

This is a fundamental point of view that has skewed our understanding of consciousness, because it rejects any claim that consciousness may exist independently of the body. The major religions and spiritual movements of the world insist that something essential survives the death of the body; call it "soul" or "spirit" or any one of a number of other terms, the message is clear: death is not the end. The

149

"ghost in the machine" may leave the machine and survive in some other state, even in some other machine.

We have a hard time understanding this idea because we have been brought up to believe that nature exists "out there" and that we are examining it, analyzing it, describing it, but all the while holding it at a distance. We are in the world, but not of it. We don't understand that our perceptions of the world are part of the world, and cannot be extricated from it.

Werner Heisenberg, one of the founders of quantum mechanics and author of the famous Uncertainty Principle, once wrote:

> Natural science does not simply describe and explain nature, it is a part of the interplay between nature and ourselves; it describes nature as exposed to our method of questioning.[45]

The conventional method of questioning so far has been to exclude phenomena that do not fit the materialist perspective. The "hard problem" of consciousness is precisely the realm of phenomena that doesn't fit. We observe phenomena that have no scientific explanation and then discard the observations because they don't fit the theory, rather than retooling the theory to fit the observations. In that way, science itself has become a kind of religion as well as a system of government that decides what is and is not "real" or part of the "realm." That science has become politicized is not news to anyone; it has had to become politicized in order to survive, for "doing science" in the twenty-first century means depending on government grants that can be awarded or withheld based on the nature of the project. That results in a limiting of the scope of scientific inquiry to satisfy political requirements.

Since doing science is expensive, there doesn't seem to be a way out of this conundrum unless theoretical physicists go the way of poets and artists in 1920s Paris, living in unheated garrets and writing calculations on the back of café napkins between glasses of absinthe. But if you're doing work with the CERN super-collider, or you need a laboratory for your genetic experiments, that solution won't cut it.

However, by showing how your work will contribute to the defense industry in some way and providing a national security rationale, the chances of being awarded a grant became proportionally greater. And it was this incestuous connection between government funding, the military-industrial complex, and science that led us down the path of either regarding the UFO Phenomenon as a military and intelligence matter or ignoring it as inconsequential if it does not pose an immediate threat.

But what of the Visitors themselves?

If the Phenomenon *is* other-worldly, and under intelligent control, then we must begin asking the hard questions for which we still do not have the answers. Have the Visitors solved the "mind-body problem"? Do they even have consciousness the way we understand the term, or are they solely intelligent the way a computer is intelligent? If we analyze the statements of observers, of close encounter witnesses, do we get closer to an understanding not only of the Visitors but of our own composition?

If the Visitors are conscious, but not human, that realization alone should considerably advance our understanding of consciousness and its relation to the human brain and nervous system. But if the Visitors are intelligent or super-intelligent machines that do not exhibit any signs of consciousness, we may have our answer after all.

As we have indicated, there is a war between our genes and our consciousness. We are not on the same page. Whatever created us, and for whatever reason, may not have taken into account the possibility that we would develop a will independent from our genetic imperative. They had planned for instincts—the instinct to survive, the instinct to reproduce—but those are not the same thing. Instincts are those programs that run the machines (the robots, the cyborgs) that are us. They are unconscious motivators, set in motion eons ago and encoded in our genes. The existence of the human will implies an ability to transcend those instincts, to suppress them. It's also what makes us aware that there may be something inherently wrong with our situation on Earth.

It's the ghost in the sekret machine.

SECTION TWO

▼

CONSCIOUSNESS

INTRODUCTION TO SECTION TWO

Gen often said that what we perceive to be coincidences are in fact carefully placed tiles in a mosaic pattern the rest of which we can't apprehend. Now Micky sensed that intricate mosaic, vast and panoramic, and mysterious.

– Dean Koontz[46]

C ONSCIOUSNESS IS THE ONE GREAT REMAINING MYSTERY for science. It's the "black box" at the heart of the human experience, the target of so many theories, ranging from the purely reductionist and mechanical to the expansive and mystical. Is consciousness "an emerging property of the brain," or is consciousness everywhere in the universe? Are animals conscious? Are machines conscious?

Are aliens conscious?

We will look at some of the most relevant theories (there are way too many to examine each one thoroughly) and suggest ways in which the study of consciousness has applications for the study of the Phenomenon.

The study of consciousness is foundational to the study of the Phenomenon, as scientists ranging from Jacques Vallée to Hal Puthoff have assured us. Actually, this should be obvious to anyone who has studied or researched Ufology for any length of time. The consciousness effects are the first ones to be experienced, from those simply witnessing the overflight of a UFO on a dark summer's night to those who find themselves in the presence of the Visitors. The experiencers themselves live through events that are almost impossible to describe in everyday terms; their psyches have been affected, and in many cases—such as those of the abductees studied by John Mack—they suffer from post-traumatic stress disorder.

The experience of the Other challenges our assumptions, not only of the Visitors—Who are they? What are they? What do they want?—but of what it means to be *us*, to be human. We need to take a good, long look at ourselves and the way our minds work. In order to approach this demanding discipline it is necessary to look at the physical structure of the brain as well as to consider the possibility that consciousness arises from the neurons . . . or maybe the microtubules . . . or perhaps the DNA molecule . . . or even from quantum effects at the smallest possible level—the Planck scale—of the material world, the level of particles and waves.

In addition, we will consider how consciousness evolved, and when it first appeared on Earth. This may help us understand how consciousness might have evolved on other planets, which might help us understand the psyches of the Others.

Taken together, these various theories and data points will contribute to a greater understanding of the enormity of the task we have set before us: *to understand the Phenomenon before we even understand ourselves.*

DNA CONSCIOUSNESS

Does genius emerge from the genes alone?
Does the largely unknown chemistry of the brain
contain at least part of the secret? . . . Or do we
ordinary men carry it irretrievably locked within
our subconscious minds?

– Loren Eiseley, *The Night Country*

ARLIER WE SAW HOW THERE MIGHT HAVE BEEN AN *unconscious* awareness of the structure of the DNA helix and the genetic code represented in the divination systems of cultures as diverse as the Yoruba tradition of Africa, European geomancy, and Chinese philosophy and mysticism. We also saw that a Nobel Prize–winning scientist, Francis Crick, held the conviction that the DNA molecule was seeded onto this planet deliberately. We also saw two Russian mathematicians discover that the structure of the genetic code itself implies an intelligence at work in its design.

Yet, we also came to the realization that DNA does not have an "immortality gene" built into it; indeed, in order for the DNA molecule to survive and thrive, it must "kill off" each generation so that a new generation—with new combinations and new capabilities—can take its place. This would make sense if a race of intelligent beings had decided that the last thing they needed was to generate billions of immortal machines. (Imagine what the earth would have looked like if the Model-T Ford had been designed to be immortal, and could reproduce endlessly!) After all, this is what we do. We invent newer and better tools, newer and better machines that are developments of the previous generations, and then jettison the older machines when the new ones come along. How many of us still use the Radio Shack TRS-80 personal computer? Or the Commodore 64? Yet the "DNA" of these older machines is built into the newer ones. The technology can be traced back to the first machines, just as our DNA can be traced back down the evolutionary line to the first living things.

However, this sets us up for a major disconnect between human beings and their environment, for human beings want to live forever (or, at least, to have the option of living for as long as they like). This desire is completely at odds with the genetic program, as far as we understand it, but it is nonetheless very real.

There have been practices and procedures designed to prolong life, whether Chinese alchemy or Indian yoga or the *elixir vitae* of the European savants, leading up to the present-day obsession with longevity, cryogenics, and the like. This is a metaphorical slap in the face of Richard Dawkins' idea of the "selfish gene," which does not want any such thing. Is it possible that this *desire* for immortality is an indication that some part of us *is* immortal?

From where did we humans get this obsession with immortality? Does the DNA molecule have a secret message encoded within its enormously long chain of nucleotides, a message that can only be decoded when humanity has reached a certain level of genetic diversity and strength? Did previous generations have to die off, as controlled by the DNA molecule, until a certain critical mass was reached, at which time the human race would be prepared to achieve (or, perhaps more correctly, *realize*) its own immortality? Is there a link between human conceptions of immortality and consciousness itself?

Is DNA the source of this consciousness? Further, if it is, is DNA *itself* conscious?

▼   ▼   ▼

There are two possible approaches to this problem. Either consciousness is built into the DNA molecule, and our desire for long life is an artifact of that molecule for reasons we don't understand, or consciousness comes from outside the system. Does our brain—if that is where consciousness is "located"—create its own signals, or does it receive signals from elsewhere? Or, possibly, both?

The aspect of the Phenomenon known as "alien abduction" provides many examples of the human nervous system acting as a kind of receiver for signals emitted by the "aliens." Alien abductees are almost unanimous in reporting that communication with their abductors is usually restricted to a kind of mental telepathy. Sometimes the communication takes the form of words, and at other times of images. Why should we take these accounts seriously?

In the first place, there is a substantial amount of consistency among the various accounts, which would seem to indicate that the abductees are witnesses to something real. Some critics have tried to invalidate those experiences with scientific-sounding reasons, such as sleep paralysis, hallucinations, etc., but those explanations would not account for the similarity in detail between abductees from various places in the world at different times, even in different decades.

The abductions take place in a state where the abductees have no control over what is happening. They are usually abducted from their homes, but at times from moving vehicles. Communication with their abductors takes place telepathically; that is, they hear or somehow sense the words "uttered" by their abductors in their heads, not through their ears.

In fact, the abductors do not have the same set of sensory apparatus that humans have. No ears are discernible, no nostrils to speak of. They have a slit for a mouth, and sometimes another slit or slits where the nose would be. Feet are virtually invisible, either due to special footwear or for other reasons. There is a vague anatomical similarity to humans (four limbs, torso, head)—which may be evidence of a shared genetic and/or environmental heritage—but not enough to indicate a common humanity.

In the United States, the abductors often take the appearance of the famous Grays: large heads, large black eyes, long tapering fingers, etc. In Europe, Africa, and Asia, however, although Grays are seen in some cases, we seem to see greater variety in appearance, although there is always general agreement that the Visitors are not human. All of this is valuable information, because it may indicate that the abductees are projecting some kind of idea or image onto their abductors, or that the abductors are manipulating the senses of the abductees for a specific purpose. That purpose may be as simple as "divide and conquer": if humans are not in agreement as to the size, shape, and characteristics of the abductors, they are not able to formulate any kind of coherent response or threat assessment across national boundaries. In other words, the abductors may be experts in propaganda and psychological warfare on a level that exceeds anything we have invented, and by using different "forms" for different countries they could intend to pit those countries against each other when the time comes (assuming we can impute a human agenda to the abductors, which is something we should not take for granted).

Even in those foreign scenarios, however, there are consistencies with their American counterparts where telepathic communication is concerned. This seems to be a characteristic that the abductors have not transcended, or perhaps they do not feel the need to do so.

Thus, we have a few features on which most "alien abduction" scenarios are based: the abductors are not human; they are associated with UFOs/UAPs; they communicate only telepathically; and they abduct human beings for purposes we do not yet understand but that often seem to have a reproductive or genetic component.

A great deal has been speculated about the motivations for what can only be described as terribly invasive and at times gruesome physical manipulations of human beings by the abductors. As there seems to be an obsession on the part of the abductors where human reproduction is concerned—with genitals, with ovaries and testes, with semen, with fertilization and conception—many critics ridicule the experiences on the assumption that if the abductors are a space-faring species with the kind of advanced technology they would need to travel enormous distances through the cosmos, they hardly could be expected to be confused or fascinated by sex. The problem with that approach is that it requires too many assumptions, none of which are made by the abductees themselves.

The first assumption is that the abductors have traveled vast distances through space. There is no evidence to support that view.

The second assumption is that they use the same, or almost the same, reproductive processes that we do (or that they reproduce at all).

The third assumption is that the abductors are organic beings, and not some type of machine.

And the fourth, and most dangerous, assumption is that their goals, motivations, and agendas can be understood in human terms.

Thus, in order to get a grasp on the problem, we have to go back to the source. We have to try to understand what human consciousness is, how it developed, and, most important, where it is going. We need desperately to understand our own nature before we can draw comparisons with the nature of beings who clearly are not like us at all. And since the consensus seems to be that the abductors—the "aliens," the "Visitors"—communicate telepathically, that means we have to take a very close look at consciousness and at our own attempts at qualifying and quantifying telepathy, paranormal abilities, and the like.

If DNA (or its predecessor, RNA) was, indeed, seeded onto this planet by another, intelligent source, as we discussed earlier, it behooves us to begin at the genetic level to gain an understanding of where consciousness came from.

The Neurobiological Theory

> It is remarkable that most of the work in both cognitive science and the neurosciences makes no reference to consciousness (or "awareness"), especially as many would regard consciousness as the major puzzle confronting the neural view of the mind. Indeed, at the present time it appears deeply mysterious to many people.
>
> —Francis Crick and Christof Koch[47]

Francis Crick himself was interested in this question, and co-authored a paper with Christof Koch on the subject. Titled "Towards a neurobiological theory of consciousness," it was published in *Seminars in the Neurosciences* in 1990. Crick believed that consciousness had to have a physical, biological origin and that the brain was the most likely locus for consciousness: what they called "neural correlates of consciousness," or NCC. To Crick, the brain was a machine that could be tinkered with, broken down, and studied with components in isolation from each other—hippocampus, thalamus, neocortex, etc.—to discover how consciousness was constructed and what processes were involved. Crick had already addressed the structure of the DNA helix and had proposed directed panspermia as the source of all DNA on the planet; now he was on the trail of the single most elusive scientific entity: the "black box" of consciousness.

A "black box" in this sense is any system in which information goes in and information comes out, but we don't know how that happens. For many of us, our computers, tablets, and smartphones are black boxes, but there are engineers and technicians for whom these devices are transparent. Consciousness, however, resists every attempt at analysis. The materialists ignore the psychologists, and both ignore the mystics.

To begin the scientific attempt to understand consciousness, Crick and Koch proposed to isolate one of the sensory systems—the visual—and concentrate on that. How does the brain respond to visual stimuli? When we see hundreds or thousands of discrete images at a time—just looking at a tree full of individual leaves, or a busy street scene in a crowded city—how do our eyes select the images we see, building a *mise en scène* in microseconds, and conveying that information to the brain? How does consciousness build that picture in an instant, only to replace it immediately with another one? When we shift our focus from looking out of a window at the sky to looking indoors at the dinner table, *what* happens in consciousness? *How* does that happen? And *where* does that happen?

> There is general agreement that we are not conscious
> of all the processes going on in our heads, though
> exactly which might be a matter of dispute.[48]

This is a fact of which the ancients were well aware. Those who study yoga, for instance, know that there are neurobiological systems operating below conscious awareness—such as peristalsis, respiration, etc., that are controlled by centers in the brain—and that techniques were developed to extend conscious control over those systems. The easiest (or, at least, the most accessible) of those systems is respiration, which can be controlled consciously through *pranayama*, or breathing exercises. Any child knows how to hold their breath until they turn blue! The practice of *pranayama* is the pathway toward conscious control of other autonomic nervous system functions, such as heart rate.

What Crick and Koch are saying, however, is that there are systems operating at deeper levels that organize visual and other sensory stimuli into the complete package we know as conscious awareness. I see you, I hear you speaking, I feel hot or cold, I smell cookies in the oven . . . all simultaneously, forming a picture of a brief moment of time. Your facial expression, combined with what you are saying—visual plus auditory events—create a global experience of the event that my senses receive and somehow organize into something coherent and comprehensible. The more complex the nervous system—the more "sophisticated" it is when it comes to discerning different colors, different sounds, different smells, etc.—the more "conscious" one becomes.

> It is probable, though, that consciousness correlates
> to some extent with the degree of complexity of any
> nervous system.[49]

This may be the crux of the problem. The more complex the system, the more complex the consciousness. And this includes all the senses, distributed throughout the nervous system. Thus, when projecting these ideas onto the typical "alien abduction" scenario, we are faced with the realization that we have no evidence that the abductors have the same sensory apparatus we do, which can have implications concerning the degree of complexity of their nervous system and hence of the abductors' type, depth, and range of consciousness.

As noted previously, ears on the "typical" Gray alien are conspicuously absent. Mouths do not seem to be used for communication. Even the nose seems vestigial, at best. If the senses are required for the development of a complex nervous system and thereby for higher consciousness, the abductors actually are lacking in that regard. That would seem to indicate—if Crick and Koch are correct in their assumptions—that the abductors either are not as conscious as we are, or have a different form of consciousness altogether, one that does not depend on the classical five senses. If that is true, the key to understanding their preoccupation with human sexuality might be right in front of us.

But before we go too far, let's see what else Crick and Koch have to teach us, and proceed from there to some current scientific speculation (and discoveries) where DNA consciousness is concerned.

First, Crick and Koch do not think that language is essential for consciousness. To support this contention they cite the examples of the "higher mammals," which "possess some of the essential features of consciousness, but not necessarily all."[50] In other words, a dog may be conscious even though a dog does not communicate in language (as we understand it), leading to the conclusion that consciousness does not equal language ability. Can we assume, however, that language ability equals consciousness? We are again in that dangerous area where machines may communicate in language but not be conscious.

René Descartes, the famous French philosopher who gave us the dictum "I think, therefore I am," had this to say about language ability:

Don't confuse speech with the natural movements that are evidence of passions and can be imitated by machines as well as by animals. And don't think, as some of the ancients did, that the beasts speak a language that we don't understand! For if that were true, then since they have many organs that are analogous to ours, they could make themselves understood by us as well as by their fellows. It is another remarkable fact that although many animals show more skill than we do in some of their actions, yet the same animals show no skill at all in plenty of others; so what they do better doesn't prove that they have minds, for if it did, they would have better minds than any of us and would out-perform us in everything. It shows rather that they don't have minds at all, and that it is nature that acts in them according to the disposition of their organs. Similarly, we with all our skill can't count the hours and measure time as accurately as a clock consisting only of wheels and springs![51]

If one were to apply Descartes' sentiment to what abductees report concerning their interactions with the Visitors, we would arrive at some interesting conclusions and be closer than ever to characterizing them as machines imitating humans.

Crick and Koch, however, go on to state:

From this it follows that a language system (of the type found in humans) is not essential for consciousness. That is, one can have the key features of consciousness without language.[52]

They then go on to make a very interesting statement regarding self-consciousness:

We shall assume that self-consciousness, that is the self-referential aspect of consciousness, is merely a special case of consciousness and is better left on one side for the moment. Volition and intentionality will also be disregarded and also various rather unusual states, such as the hypnotic state, lucid dreaming and sleep walking, unless they turn out to have special features that make them experimentally advantageous.[53]

These are, of course, the very states that compel us to investigate consciousness in the first place, the self-referential aspect or self-consciousness paramount among them. If you propose a figure that does not have language as we know it and is not equipped with self-consciousness or self-referential consciousness, then you are halfway toward describing the classical "alien." According to the alien abductee reports in David Jacobs, Mack, Bryan, and others, the alien abductor is without much in the way of personal identification. There is no differentiation between one "alien" and another in terms of facial features, clothing, speech patterns, gestures, etc. This would imply a level of self-awareness or self-referential consciousness (the concept of a self that is distinguishable from the group, an independent being) that is minimal, at best. Of course, language ability is another deficit when it comes to interaction with the alien abductors. One may as well be dealing with machines.

In the very early days of artificial intelligence (AI) research, a program known as ELIZA was developed. It was created in the 1960s as perhaps the earliest version of a "chatterbot": a program that simulated human speech. One could ask ELIZA questions and receive answers that often seemed appropriate and relevant, as if speaking to a human being on the other side of the computer screen. It is clunky compared to present-day systems, but it is amazing to contemplate the degree to which ELIZA was considered to be cutting edge at the time.

What, then, if that which is perceived as "alien" communication is nothing other than a more refined version of ELIZA, i.e., a machine-generated communication system that is based on images rather than sounds? Would not the especially large eyes of the "aliens" seem to reinforce the idea that images are their natural method of communication? And the lack of complete reciprocal verbal exchanges with the "aliens" would seem to imply a level of pre-programming rather than a natural flow of language, concepts, and repartee. It just may be that what we understand as an advanced species is not quite as advanced as they would have us believe.

Or, they are not a species at all.

▼   ▼   ▼

If language is not necessary to consciousness, however, then there is a possibility that the Alien is conscious; just not very verbal. Crick and Koch propose that the most accessible path to understanding consciousness is through the visual sense rather than the aural, as discussed earlier. (This was a path taken by Descartes, as well, who wrote a groundbreaking paper on optics.) They argue that if the neural pathways that control vision in the brain can be mapped, analyzed, and understood, we are partway toward grasping the essential nature of consciousness.

This is because the visual field is enormous and consists of a wide variety of shapes, colors, dimensions, and movement. The "processing power" required to look at a street scene or a painting or just your living room is considerable; how much more so the prioritizing of what is seen, the assignment of "values" to the visible objects, the organization of everything seen into a coherent whole that may last for no longer than a nanosecond. When we look at a scene before us, we are not completely consciously aware of every object in that scene. If I see a row of books, for instance, I know it is a row of books and I may even have the immediate sense that they are hardcovers or paperbacks or some combination of these. I will not be aware immediately of the titles on the spines of the books. That requires a different sort of attention, one calibrated toward language—the languages used on the spines—and to the linearity in time required to read each one. You can't read all the spines at once.[54]

However, your memory may have registered all of the titles; they're just not easily accessible. In addition, there is short-term memory and long-term memory and perhaps shades of difference between them. This makes image capture and retrieval even more problematic . . . for a human, but not for a machine. The machine can record and store all the visual data, but it is still up to a "program" to decide what data should be retrieved at any given time.

When it comes to short-term memory, the prognosis seems rather more extreme:

> No case of a person who is conscious but has lost all forms of short-term memory has been reported.[55]

It *would* seem rather bizarre to have a person who is conscious, yet only has long-term memory and no short-term memory at all. They might appear to be either unconscious or simply not present. But what of a patient who had only short-term memory but could not form long-term memories? The famous amnesic patient known as "H.M." was fully conscious, even though sections of his brain had been removed to ease his suffering from epilepsy. He had tremendous difficulty in forming long-term memories, but his ability at forming short-term memories remained intact.[56]

Then we have the fact that blind people—people born sightless—are obviously wholly conscious. So are people who cannot hear. Or speak. Helen Keller was conscious, maybe more than most. And we know all of this because those people can communicate. With or without language as we understand it, they communicate.

Thus, is communication with an "alien" species analogous to communication with a human being that is challenged in a sensory way? Or is it the other way around: does the alien species feel that humans are sense-challenged, and have resorted to pictures to communicate with us because that is the way *we* communicate with children who have not yet learned to read?

▼   ▼   ▼

Crick and Koch refer to a concept known as "binding." This is a way of describing what happens when we are speaking with another person and we are aware of their appearance, the words they are forming, and other ancillary data that are essential to carrying out our end of the conversation. These various senses—sight, hearing, and so forth—have to "bind" together to create the overall experience.

Then they amplify this concept a bit further by saying there are three types of binding. The first, they claim, is "probably determined by genes and by developmental processes that have evolved due to the experience of our distant ancestors."[57] This is the first time that these two scientists have acknowledged the genetic component of an essential aspect of consciousness: binding together data from the various senses into a coherent whole. Further, this aspect of consciousness may also have been developed over the eons as we proceeded along the evolutionary path.

The second type of binding is one with which we are all familiar: learning. They use the example of learning the letters of the alphabet as representative of this type of binding. Of course, the alphabet is visual but it also represents sounds. The combination of letters represents an even more advanced type of learning, which is the association of words with concepts: nouns, verbs, etc. Learning also obviously involves those parts of the brain that concern memory.

The third and final type of binding, according to Crick and Koch, has nothing to do with the first two. This type is not related to a genetic endowment, nor to repetitive learning, but "applies particularly to objects whose exact combination of features may be quite novel to us. The neurons actively involved are unlikely all to be strongly connected together."[58]

Neurons are connected to each other through use, i.e., through repetitive learning or through some genetic mechanism with which we are as yet unfamiliar. However, there are times when we experience something for which we have no existing neural connection, no combination of nerves and synapses that form the appropriate pattern in our brain.

What happens then?

Unless the experience is repeated enough times, this type of event will not translate into a neural pattern that can pass into memory, to be accessed again at a later time as something familiar and related to other categories in memory. However, the event *is* recorded through what Crick and Koch call a "serial attentional mechanism." The fact of the event is recorded, but there is insufficient binding of the details of the event to produce an intelligible memory, since it represents a novel pattern that had not been experienced previously. This is a fancy way of stating the obvious: events that capture our attention are recorded, even if they have no precursor:

> In other words, *that awareness and attention are intimately bound together.*[59]

They do not seem to acknowledge the possibility that consciousness has an interpretive element: that the unfamiliar, "novel" object may be subject to an attempt by the consciousness to interpret what it sees, by associating it (rightly or wrongly) with something else in memory.

An adage that is often used in the law enforcement community is that eyewitness testimony is notoriously unreliable. Ten different people can witness an event, resulting in ten different accounts that contradict each other. However, there is usually consensus among all eyewitnesses that the event in question did take place: a robbery, a murder, etc. The differences are in the details: was the assailant tall or short, wearing a blue jacket or a green one, etc.? Quite often discrepancies between eyewitness accounts of a UFO sighting are used to discredit the sighting for that very reason, but without acknowledging that the event actually took place. In the case of a murder there is a corpse, and the only problem is the number of contradictions between eyewitnesses as to exactly how the murder occurred. In the case of a UFO sighting, the event itself is discredited. In other words, there is no corpse.

If the "aliens" looked like us, and landed in an aircraft, the number of correlations between the different eyewitness sightings would be greater. That is because we know what *we* look like, and we know what an aircraft looks like. There may be discrepancies in how we remember the event, but they would be minor relative to the overall impression of an aircraft landing and people emerging.

But if the "aliens" don't look like us, and their "craft" is not like anything we have ever seen, we are in unmapped territory. Our senses will struggle to interpret what we perceive and will associate various aspects of what we see with previously recorded memories of objects similar in appearance. Unfortunately, that will account for a much greater degree of discrepancy between eyewitnesses.

Critics point to the fact that since Kenneth Arnold—in his famous UFO sighting of 1947 in the Cascades—referred to the alien craft as "flying saucers," eyewitnesses began seeing "flying saucers" as if inspired by Arnold. In reality, Arnold did not describe the UFOs he saw as saucers. His drawings of the craft made them appear more like the flying wing aircraft that had been in development in different countries over the prior decade. Yet, he referred to their aeronautical characteristics as looking like saucers skipping over the water. This gave birth to the term "flying saucer" and to the image of a circular object—shaped like a saucer—rather than to a comparison to other geometric figures that have since become more familiar, such as the cigar-shaped UFO.

The damage had been done, however, at least in the eyes of the critics. Since Arnold seemingly had described saucers, well, people saw saucers. This was considered evidence of their gullibility or their duplicity, which is missing the point. Various critics have maintained that our impression of alien craft owes more to tabloid newspapers, pulp magazines, and Hollywood than it does to anything "real."

But what the witnesses to UFO events see is a phenomenon with which they have had no previous personal experience. They can be said to "project" ideas about what the UFO *should be* onto what they are actually seeing, and they derive these ideas from the media. In other cases, witnesses who have not been quite so prepared by media reports and movies about aliens and UFOs report seeing different shapes behaving in different ways. In every case, however, the event is something that defies any previously held memory of human experience, and that is pretty much the point. Awareness and attention, as Crick and Koch have written, are intimately bound together. Where the scientists were referring to events that take place along the neural networks in nanoseconds, we can extrapolate their assumptions to the macro level of UFO sightings (and other "paranormal" events). Most human beings have no memory of witnessing a UFO event; when they do witness one, they do not have the neurobiological context within which to "see" it and recognize it. So a kind of myth builds up around the event, and it binds with other myths that have evolved similarly (sightings by other people, for instance), and the memory of the initial encounter becomes "homogenized" with those of other individuals.

While there is *awareness* of the UFO event, *attention* may in some cases be lacking. How does one focus one's attention on something that defies immediate identification?

One is reminded of the old television series, *Superman*, and the tag line: "It's a bird! It's a plane! It's Superman!" The immediate natural reaction to seeing a human being fly unaided through the air, according to those who wrote the tag line, was that it must be a bird. We know that birds fly, so that was a safe assumption until the "bird" exhibited peculiar aerial characteristics and had no wings that would explain how it was aloft. The next assumption therefore had to be that it was a plane. Birds and planes are things that we know can fly. But the observation did not match either a bird or a plane. It was a human—or a human-like being—flying through the air without wings, without a machine. That was the only possible conclusion, though it took a while to get there and it defied rationality.

Thus it is with UFO sightings and human consciousness.

▼   ▼   ▼

In the end, Crick and Koch are unable to say anything definitive about consciousness. In fact, all they can do is suggest avenues for further experimentation and research. They feel that the visual field has the most potential for this type of approach, which is why they spend so much time on it. But a few gems do manage to survive the frustration and impatience of these two scientists, and they are what have brought us to this place. The idea that awareness and attention are different yet both are necessary for an understanding of the basic mechanisms of consciousness is one of them. The idea that our neural networks—a legacy of genes and evolution—may not be up to the task of recognizing certain stimuli is also important.

Conscious Machines?

Crick would later publish a book[60] suggesting that consciousness should be the focus of modern scientific inquiry. However, his approach was a materialist one that was based on the building blocks of neurons, cells, atoms, and molecules and would fall within the purview of classical physics and neurobiology. Crick and, later, his colleague Christof Koch would be adamant that consciousness was an artifact of the purely biochemical evolutionary process, an emergent property of the brain. There was no quantum mechanical aspect, they would claim, nor would machines ever be conscious.

While this is presented as a scientific view, they remain unable to offer any kind of workable, testable hypothesis for the claim that consciousness is nothing more than a phenomenon comprising massive numbers of brain cells firing in various patterns, reflecting responses to sensory stimuli. The position of Crick, Koch, et al. is a kind of ideology in its own right. They continued to assert that a study of the processes involved in vision would present the best avenue for discovering the nature and origin of human consciousness, but without giving any evidence for why that should be. The problem remains that there is no generally accepted definition of what it means to be conscious.

This has led to a debate among scientists as to whether machines would or could eventually become conscious. In the absence of any clear definition of consciousness it is as valid to claim that machines could become conscious as to claim the opposite: that machines are machines, and humans are humans, and consciousness is a characteristic of human beings alone. Those who claim that machines could become conscious are those who have adopted the attitude that computers—because they calculate, act intelligently, and can be made to mimic human behavior—are somehow analogous to human beings and capable of acquiring human consciousness. Those who insist that machines will never become conscious point to the computer analogy as a dead end leading nowhere because there are no precise equivalencies between the memory system of a computer (for instance) and that of a human being. While it is common to state that humans store memory in their brains, that is not exactly true. There is no one cell or group of cells that can be identified as the place where a specific memory is "stored." There is no cell corresponding to the image of a house, for instance, or the smell of steamed rice. If we cannot identify a physical location in the brain for a memory, we are not like computers at all; we know precisely where a computer's memories are stored.

Both points of view avoid the central question, however, and are based on a number of assumptions that so far have not been proven to anyone's satisfaction. Is consciousness simply the ability to receive and process sensory information, remember it, build on memories, and react accordingly? If so, then the comparison to computers is apt. But human beings do not act according to logical rules. They are messy, willful, inconsistent in their behavior, hold untenable beliefs, get angry, fall in love, bond with their children, and often act against their own best interests. Human beings are emotional.

Human emotions come from a variety of sources, but one thing is certain: every human being has a history. Each of us has evolved from a fetus to a full-grown organism, and during that period of growth and development we have accumulated experiences that have molded us into the individual persons we are. Our emotions, our thoughts, our memories, our culture and how we respond to it—all are unique to each of us.

A machine just *is*. It is built according to specifications. Its memory is composed of data we input. It has a specific purpose and function. Its qualities and capabilities arrive full blown from the head (or heads) of whoever designed it. It has no history, no inner developmental process. It has no will. It cannot lust. It cannot love. It has no reproductive urge. It does not gestate. It does not get hungry or thirsty. It does not know fear or anxiety. It is not aware of its own impending demise. It does not fear other machines. It does not love other machines.

It does not fear or love *us*.

In order for AI to succeed the way the neurobiologists and scientists want it to, a way must be found to give a machine—a computer, a robot, an android—a subconscious. This subconscious must be equipped with an algorithm that enables it to behave seemingly capriciously, illogically. It has to be a place where random thoughts and experiences are stored, only to be recovered in moments of stress or in sleep.

In other words, the machine—computer, robot, android—must be able to *dream*.

Until that problem is solved, all the theorizing about the roles the cerebrum, cerebellum, amygdala, neocortex, prefrontal cortex, limbic system, etc., play in consciousness will come to naught. Human consciousness is much more than a mechanism for interpreting sensory stimulation. It is the product of eons of evolution, both genetic and social. The two main groups fighting over whether or not a machine may become conscious—or whether we may one day create a conscious machine—are both wrong, since neither side has defined consciousness successfully.

We will, however, create machines that can think; that follow orders; that move independently; that even have certain social rules and roles embedded in their operating systems. We may develop them for long space flights, as replacements for human beings who would not survive a hundred years or more in suspended animation aboard a starship. They would have just enough information encoded in their processors to enable them to function intelligently in flying the starship (just another machine, after all), landing on a planet, and leaving the ship to collect samples of chemical and biological importance, and communicate the results back to us on Earth.

You know—just as it is said the "aliens" are doing right here, right now.

▼   ▼   ▼

While all of these analogies are suggestive, we really have to do our best to involve current scientific investigation and inquiry in order to determine how consciousness first appeared on our planet, or at least in our species. Knowing something of the timeline may give us an idea of where consciousness came from, and of what it may be composed. One of the theoreticians who believes that we can find the original traces of consciousness in our DNA is John Grandy.

Building on the work of Crick and Koch, Grandy takes their concept of "neural correlates of consciousness" (NCC) a step further. He posits the existence of "neuro*genetic* correlates of consciousness" (NgCC), which "focuses on the study of genes and gene products (e.g., transcription factors) that are involved in (and have an effect on) the conscious experience."[61] He calls this the "*continuum of neuron-based consciousness*."[62] The idea is that there exists a genetic layer operating below the neuron-based layer. He finds support for this position in the fact of neurodegeneration (such as in Alzheimer's disease, one of Grandy's major research areas), in which a decrease or degeneration of conscious faculties is accompanied if not caused by genetic deterioration.

The genes do not stop functioning once the body has been formed. Since the genes remain active—some more than others, and some still active days or even weeks after death—it seems reasonable to suppose that they have an effect on consciousness. The genes create the neurons, which operate on a more macro level than the genes; both affect consciousness.

Grandy even goes so far as to suggest that there are quantum effects at the DNA scale, citing hydrogen bonding forces between the nucleotides, for instance.[63] This interaction of DNA nucleotides and proteins contributes to what he calls "DNA consciousness," which can "give rise to higher degrees of consciousness, e.g., cellular consciousness and human consciousness."[64]

To review his theory, there is a quantum effect taking place at the smallest—the genomic—scale of the neuron, which then gives rise to DNA consciousness, which gives rise to higher order neurogenetic consciousness and eventually to human consciousness. The DNA molecule is the locus for a quantum effect, which would seem to imply that the brain is what some theorists have proposed it is: a quantum computer.

Grandy first introduced these concepts at a conference in Sweden in 2011, and has expanded and refined the concepts considerably since then. His ideas seem reasonable, since he associates psychiatric disorders (and hence disorders of consciousness) with genetic mutations. This would imply, logically, that consciousness has a genetic basis.

The problem remains, however: What *is* consciousness? A person with Alzheimer's disease (AD) is conscious at least in the sense of being awake; there may be a disconnect between awareness of one's immediate surroundings and what one is experiencing at the time, or a sporadic and often transient loss of memory that is the hallmark of the disease, but is that a fault to be located in consciousness itself, or in the medium through which we experience consciousness? If our radio is damaged, does that mean the station has stopped broadcasting? If there is a physical disorder in the brain or damage to the nervous system, does that mean that consciousness has left the building? Are we only confusing *awareness* with consciousness? A person with AD has *attention*; it's just not directed at something we can all agree is relevant or extant. The person with AD may not be aware of their surroundings from time to time (a situation that gradually worsens), but they are aware of something. The binding that Crick and Koch reference may have broken down: awareness and attention have either become separated in the case of AD, or they have bonded to some other set of stimuli.

The heartbreaking symptoms of AD, with which everyone is familiar, include the inability of a patient to recognize their child or spouse; to believe that a dead relative is still alive; and forgetting where things are located, or even what they are called. This indicates a breakdown in the retrieval of memories, short term as well as long term, but not in the breakdown of consciousness itself. The person suffering from AD becomes agitated or depressed, which is an understandable reaction to the experience of dealing with a faulty memory as well as the frustration of having a conversation and forgetting basic vocabulary. To use our other metaphor, the patient is having difficulty finding the right radio station because the tuning dial is broken.

Physically, the brain begins shrinking. Cell death at the neural level takes place. The awareness of AD patients begins shrinking as well, as they become more isolated from society and more emotionally withdrawn. Eventually they become unable to care for themselves, to maintain personal hygiene, and to keep their balance. This causes them emotional distress, which implies consciousness. Can one have emotions without consciousness? Self-awareness, which is necessary for feeling distress over the gradual loss of memory and mental acuity, is one of the issues that Crick and Koch said they could not address in their research on consciousness.

In the past several years more and more discoveries about the genetic basis for AD have been made and specific genes identified as the "carriers" of the disease. This, however, is not sufficient to "prove" a genetic basis for consciousness, as not everyone with the genes actually develops AD. Other factors are involved but are so far unidentified. What if a potential AD sufferer is dealt a hand of cards, one of which is the AD card, but refuses to play it? Is there an unconscious motivation at work in the human mind that selects from an array of possible illnesses, based on some basic—possibly even essential—factor that we do not as yet recognize?

▼　　▼　　▼

Is DNA the ground of consciousness, or is there perhaps another—deeper—level that we need to explore? If we do not share DNA with the Visitors, or if our genetic complements are incompatible, then the Visitors' capability of communicating with us must take place within some other framework. That basic and essential factor may hold the key to understanding not only ourselves but the Phenomenon as well.

THE "HARD PROBLEM"

The mystical summons up the mechanical.

– Henri Bergson[65]

THE WORLD'S GREAT PHILOSOPHERS STRUGGLED WITH understanding how the physical world is made and of what basic materials it is composed. There was—and remains—a basic assumption that the world we experience through our senses and with our scientific apparatus is constructed of building blocks which, when assembled in various combinations, give rise to everything in creation. For the ancient Greeks, this meant atoms. While they had never seen an atom because they did not have the technology to do so, they postulated its existence through logic and reason. Plato expanded upon a very old idea concerning the elements and introduced us to the idea that there were four basic states of matter: fire, earth, air, and water, plus another rather more amorphous element called "aether." This assumption seems naïve today, especially in light of the Table of Elements, with which every schoolchild is familiar. Nevertheless, it was an assumption that was shared among the ancients in many cultures around the world. For the Chinese, there were five such "basic" elements or processes: fire, wood, water, earth, and metal.

It should be pointed out that these were names or placeholders for "qualia": that is, they were not meant to be taken literally but as descriptions of forces, of qualities. In Western languages, for instance, it became common usage to describe certain persons as "saturnine" or "mercurial" or "martial," which were references to qualities associated with the planets Saturn, Mercury, and Mars in astrology and mythology rather than a statement concerning that person's planet of origin. A person also could have a fiery temperament, or wooden affect, an airy personality, etc. These are qualia.

In neurobiology, the term "qualia" is taken to refer to subjective experiences of objects. The usual example given of a quale (singular of qualia) is the color "red." There are countless shades of the color red, and they are perceived differently by different persons, all of whom have different associations and therefore different experiences of the same color. We may say that light at a certain frequency presents as the color "red," but we don't know what that means for consciousness. Is it possible for a person who sees red to communicate to a blind person the experience of red? We can discuss the color from a scientific perspective, but we can't communicate the experience itself.

The four "Platonic" elements most likely began as qualia, as subjective experiences of these otherwise invisible forces. They were derived from experience, such as *fire*, which seemed to represent a set of ideas and effects rather than a discrete chemical or physical element. Fire carries with it the idea of heat, of light, of destructive force, of dryness, etc., which are all subjective experiences but which can be quantified to some degree. The same applies to the other "elements." We can see that these ideas formed the beginning of our natural sciences, and were picked up and amplified by thinkers such as Aristotle, Galen, Avicenna, and others.

But these represent the experiences of human beings on the planet Earth. Qualia on other planets or among members of nonhuman species on those planets would necessarily be different unless the other planet had an environment identical to that of Earth, and life forms had evolved on that planet identical to humans from Earth. Any slight difference in the physics or chemistry or even the topology of the planet would result in different experiences of the same effects. This idea will become important as we go along.

Eventually, modern science in the twentieth century came to recognize four forces in the universe: the electromagnetic force, the strong nuclear force, the weak nuclear force, and gravity. Gravity is the outlier, since no one really knows how gravity works with regard to the other three forces. If it could be shown that all four forces work according to some central pattern or formula, we would have what is known as the Grand Unified Theory, or GUT: a way of combining all four forces into a single force, a single statement about the nature of the universe. This would also be a way of combining general relativity with quantum mechanics: a marriage of incompatibles that is resisting all efforts by their respective matchmakers. Gravity is the force—or field—that is the culprit leading some theoretical physicists in some intriguing directions—such as string theory—in order to reconcile the differences.

▼   ▼   ▼

Because all matter can be broken down into smaller and smaller components, it is suggested that creation began from the smallest of these components and that their combinations produced materials as diverse as animals and plants and humans and planets. The problem with this approach is that one always discovers smaller and smaller components, and when they are found they do not always behave in ways that suggest they share the same universe we do. When one approaches the subatomic level to the realm of the quantum particles, mysteries begin to multiply.

Paradoxically, it very well may be that these smallest of known particles hold the key to consciousness and may even point the way toward such phenomena as mental telepathy, remote viewing, and other "psychic" manifestations. If so, that aspect of the Phenomenon that is the most mysterious and inexplicable and attracts the greatest hostility from scientific and academic circles—contact and communication with nonhuman entities, such as in close encounters and cases of so-called "alien abduction"—may be capable of being explained and even verified.

Communication with beings from planets other than Earth is believed by many scientists—and especially linguists and anthropologists—to be impossible, due to the fact that these beings would have evolved differently in their different environments and would not share any sensory apparatus in common with humans (such as eyes that would sense the same wavelengths of light that ours do, ears that would hear the same range of sound, etc., assuming that eyes, ears, and other sensory apparatus existed in their biology), thus rendering any possibility of actual communication moot. If, however, consciousness is *not* a product of the physical brain, it would not be restricted to a human nervous system or its neurobiology. The brain may be a receiver of consciousness, much like a radio receives a broadcast signal. The more sophisticated and integrated the receiver, the broader range of signals it could receive. If consciousness is universal—if, indeed, as some scientists speculate, the universe itself is conscious—then the possibility of communication with an alien species becomes conceivable. It would only require the human brain to "tune in" to the right station.

(As mentioned previously, one of the founding members of the To The Stars Academy is Hal Puthoff, award-winning physicist and the foremost pioneer in the field of remote viewing. His research along these lines—for both private companies and for the US government—has led him to study the UFO Phenomenon as well. This fact alone should be enough to imply that the Phenomenon is connected to some of the so-far-unexplained artifacts of human consciousness. In addition, other scientists affiliated with To The Stars Academy, such as geneticist Garry Nolan, are isolating and measuring observable physical effects on the human immune system found specifically in contactees.)

In the remaining chapters of this section, we will examine some of these conditions and the strategies that could be employed in any attempt at communication with nonhuman entities, using as a reference some of the more reliable accounts of "alien abduction" with a special concentration on how communication was described in these encounters. To do this, we will briefly discuss some of the latest theories concerning information theory, artificial intelligence, and quantum physics as delineating the environment within which this phenomenon may be understood, buttressed by recent discoveries in neurobiology. We will then talk about the well-documented cases of remote viewing and associated phenomena, and will finish by asking an important question: What if the ancients (and even more recent scientists and philosophers) had the mechanism right—that the various atoms and subatomic particles are the components that compose the known universe—but had the perspective wrong; that it is not bottom-up, but top-down?

The "Hard Problem" of Consciousness

No two physicists, neurobiologists, psychologists (or mystics!) can seem to agree on what constitutes consciousness, where it comes from, and when it began. No one can agree if other organic beings have consciousness. Some claim that the universe is conscious; others, that every object in the universe is conscious; still others, that it is impossible for consciousness to examine itself.

We can gain new insights into the human condition by discussing some of the dominant theories of consciousness and their implications for alien-human contact. We have a wealth of data on the experiences of abductees, collected in volumes by John Mack, David Jacobs, and others. Although scientists may scoff at these accounts and denigrate them as hoaxes, fanciful imaginings, delusions, or the ravings of lunatics, we find that the consistency in their accounts of human confrontation with the Other suggests fertile lines of inquiry into the nature of the Phenomenon. It is a symbiotic relationship, for the more we learn about human consciousness the more we can begin to understand the reality behind close encounter phenomena.

We will break down modern consciousness theories into digestible parts. We will look at chaos theory, quantum consciousness, and DNA consciousness. We will cite works by David Chalmers, Larry Vandervert, Stanislas Dehaene, Sir Roger Penrose, and others. At the same time we will show how each approach may teach us something important about alien-human contact, the possibility of alien-human communication, the degree to which the

Visitors exhibit both human and nonhuman characteristics, and how analyzing alien-human contact can suggest new avenues of discovery.

We've already discussed DNA at some length and have shown that it appears as though there has been "unconscious" awareness of the structure of the genetic code in ancient times, long before the latest scientific revolution. We will take a deeper dive into this possibility and ask if it is possible that DNA itself is conscious, or if it provides the physical substrate for consciousness. We will look at the neurobiology of the cerebrospinal system and ask if the neurons themselves are the source of human consciousness.

Building on that, we will dive even deeper, to the quantum level, and ask if quantum effects obtain at the level of the genome or at the neural level, as some have suggested. And we will see if chaos theory and information theory can help explain some of the science behind consciousness. This will not be an exhaustive examination of these approaches, of course, but even so we will introduce concepts and terminology that may seem challenging at first. However, we feel that the Phenomenon requires us to ask these questions and to suggest some answers.

THE STRUCTURE OF THE HUMAN BRAIN

Let me see. *(Takes the skull.)* Alas, poor Yorick!
I knew him, Horatio, a fellow of infinite jest, of
most excellent fancy.

   – William Shakespeare, *Hamlet*, Act 5, Scene 1

ANY CONSIDERATION OF THEORIES OF CONSCIOUSNESS require at least a working knowledge of the structure of the nervous system and the brain, including the spinal column. That is because most scientists agree that consciousness must be an artifact of the nervous system and the brain, since that is where sensory inputs originate and wind up, respectively. It is believed that consciousness requires sensory input and memory management and that these functions can be traced directly to areas of the brain. This is by no means universally accepted, of course, mostly because there is no standard definition of consciousness. However, we can observe that persons who suffer injuries to the brain or to the nervous system behave differently or find their mental and/or emotional abilities impaired or altered in some way. A blow to the head can cause a person to become "unconscious," yet we also say of someone deeply asleep that they are "unconscious." These are not the same states—one is the result of physical damage and the other is natural, and it is more difficult to "recover" from the one than the other. Thus there are problems with the terminology, and these semantic issues increase the difficulty of determining the exact nature of consciousness.

So before we go much further it is essential that we agree on some basic principles when it comes to the brain and its nervous system. This may seem like an unnecessary step, but how are we to understand alien biology—the "extraterrestrial biological entities," or EBEs—if we don't understand our own? And it may be that applying what we know of our own cerebrospinal nervous system may offer us insights into the structure and characteristics of the Visitors.

What we are calling the cerebrospinal nervous system or the central nervous system (CNS) includes the brain and the spinal cord. The peripheral nervous system (PNS) includes the nerve channels in the rest of the body: the arms, legs, torso, etc. There are two distinct systems in the PNS: the autonomic nervous system (ANS), which is in charge of the involuntary systems such as breathing, heart rate, peristalsis, etc., and the somatic nervous system (SNS), which controls voluntary movements. The ANS includes the sympathetic nervous system (sometimes called the "fight or flight" system), the parasympathetic nervous system ("rest and digest"), and the enteric nervous system (sometimes called "the second brain," but which is embedded within the gastrointestinal tract from the esophagus to the anus and can operate independently of the other two systems). While these systems are part of the peripheral nervous system, they are still regulated by operations in the CNS; for example, the thalamus (which is explained below) regulates the ANS. We will not concentrate on the systems themselves at this time, but instead shift our attention to the structure of the brain.

When one thinks of the brain, one usually conjures up an image of the *cerebrum* (Figure 11). This is the large organ that is frequently seen in episodes of *CSI* or in other television program autopsies and postmortems in which the brain is lifted out of the skull and examined. It has the appearance of a walnut and weighs about three pounds. The outside of the cerebrum is a grayish color (hence the term "gray matter"), and is known as the *cortex*. Down just a few millimeters below the surface of the cortex, the cerebrum is white in appearance.

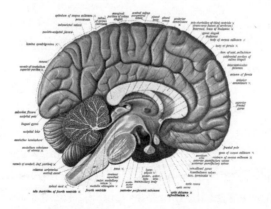

Figure 11. An old diagram of the brain, showing its structure and complexity.

The cortex—like the walnut it so closely resembles—is full of what appear to be grooves, wrinkles, and other marks. It appears random, but isn't. In fact, those grooves and wrinkles are amazingly the same from brain to brain, and this similarity is what helps researchers identify specific areas of the cortex and what they control, like a topographical map. They divide the cortex into discrete regions: the occipital lobe at the rear, the frontal lobe behind the forehead, the temporal lobe above the ear, and the parietal lobe at the top, which runs under the scalp.

The cerebrum consists of two hemispheres: the right and the left. Typically, in right-handed people, the right hemisphere controls the left side of the body and creativity, while the left hemisphere controls the right side of the body as well as logic and reason. In a sense, you could make the argument that the right hemisphere is art and the left is science, and that's useful as an analogy to a certain extent, but the reality is much more complicated than that.

Between the two hemispheres is the *corpus callosum*: a wall or bridge between the two that is concerned with cross-hemisphere communication.

Deeper within the brain are the thalamus, the hypothalamus, and the pituitary gland. The *thalamus*—a Greek word meaning "chamber"—is believed to be the center of the brain that regulates sleep and wakefulness, and therefore consciousness. It receives inputs from all the senses except the olfactory (smell). The thalamus consists of two halves, and it is in the middle of the two halves that we find the pineal gland, which some mod-ern-day mystics associate with the "third eye" and which the French philosopher René Descartes associated with vision and the soul.

The *hypothalamus* (i.e., "under the thalamus") is a gland that regulates hunger and thirst, fatigue, and body temperature. At the bottom of the hypothalamus is the pituitary gland, which secretes hormones that regulate blood pressure, metabolism, the sex organs, pregnancy and childbirth, and other systems.

The *amygdala* is an almond-shaped bundle of nuclei that is located in the temporal lobe. There are two amyg-dalae, one on the right side of the brain and one on the left. The amygdalae are associated with emotional response in humans, with one responsible for fear and the other for emotions, both positive and negative. It is believed that the amygdala processes perceived threats to the body and regulates reward/punish reactions; i.e., fear as a survival mechanism. During sleep, when most of the other brain centers are quiet, the amygdala remains alert and active. Therefore one's emotional "self" is operating, even while asleep, while one's rational "self" is shut off. Thus the inhi-bitions one has while awake are neutralized during sleep, which is why Freud considered dreams to be the "royal road to the unconscious," to the consciously suppressed material of our psyches.

The "alien abduction" scenarios seem to be almost wholly focused on the amygdala, both in terms of the dreamlike content of the experience—which is why many psychologists insist that the alien abduction experience is related to sleep disorders—as well as the intense fear aspect of the encounters. (Consider that if the dream indeed is the road to the unconscious, it is possible that the unconscious may be a road to somewhere else.)

Below the cerebrum we find the *cerebellum* (the "little cerebrum"). It has a design similar to that of the cerebrum, with folds and grooves, and a cortex as well. Its function is concerned primarily with the control of fine motor movements, such as the ability to pick up objects with accuracy. People who have damage to the cerebellum, for instance, experience tremors and shaking hands.

Connected to the cerebellum is the brain stem, which is the extension or continuation of the spinal cord. All of the nerves that carry information to the brain from the rest of the body pass up the spinal cord into the brain stem, and all the commands issued by the brain to the rest of the body are also carried down the spinal cord from the brain stem. The larger structures of the brain—the cerebrum and the cerebellum—are believed to have evolved from the brain stem and spinal cord.

The spinal cord is itself a remarkable system for handling all of the sensory inputs coming in from outside the body (cold, heat, smell, touch, taste, sounds, etc.) as well as from inside the body (everything from a stomach ache to a "phantom limb"). It is easy to see how modern-day thinkers tend to associate the central nervous system with a computer: input devices, controllers, RAM, ROM, and a CPU.

The *medulla* (or "middle") extends from the brain stem, and a body known as the *pons* (or "bridge") extends from the medulla. The pineal gland and the pituitary gland are appendages, in a sense, of the pons.

If we collect all of these systems and functions in a basket, we can see that our basket could account for everything (or almost everything) that we think of as consciousness. There are, of course, many more aspects to the brain's structure, more functions, and more capabilities, but these are some of the most important. A lot of mystical thinking over the ages has been associated with the brain and its various "departments," and the whole practice of phrenology—a way of associating specific qualities with specific areas of the brain, used as a fortune-telling tool—was predicated on the idea that more than just conscious awareness was contained within the brain: it was also the place where space and time could meet, futures could be predicted, messages received across space, love identified, dreams interpreted, and pain and death averted. So much superstitious nonsense, of course, but when we look back at it—at all those ceramic or porcelain heads with regions marked on them in black ink, as seen in Figure 12—we realize that maybe they were on to something. Their divisions of the brain were naïve and uninformed, but the core idea—that consciousness and all that implies is contained within the brain, and that different characteristics, functions, and capabilities can be located at specific sites on the cerebral cortex—has withstood the test of time.

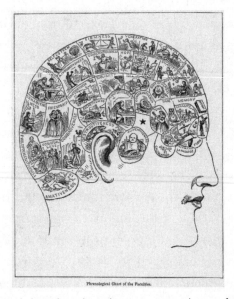

Figure 12. Typical phrenology chart, showing presumed areas of the brain that control various functions.

The problem remains, however: how do the brain and its individual components create consciousness? To use the computer analogy, we know our mouse and keyboard communicate to the computer's CPU, but we don't know how. We can see images on our computer's screen, but we really don't know how they got there. We can trace the cables, peer inside the machine, but we don't know what we're looking at. If we're good, we can identify the hard drive, maybe a CD drive, or on the older systems a floppy disk drive. We can find the motherboard. But then we come to a crashing halt. We still have not identified consciousness. We know it's in there, but not much more than that.

Time to look more closely.

The Neurons

The neurons comprise the battlefield of the conscious-ness wars. All roads lead to the neurons, whether you are talking about DNA consciousness, quantum conscious-ness, chaos theory, or any other model or hypothesis. While we know that different areas of the brain control or regulate different bodily functions, we don't know how they do it. In fact, sometimes one area of the brain seems to pick up where another left off, due to illness or physi-cal damage, for instance. In one particularly startling case, reported in *The Lancet* in 2007, a man whose brain had shrunk to 50 to 75 percent of its normal size was com-pletely conscious and still functioning in the world when it should have been impossible for him to do so. He was forty-four years old, married, and had two children and a job as a civil servant. He had had hydrocephalus (water on the brain) as an infant, and the water was drained away with a shunt that was not removed until he was a teenager. The fluid evidently built back up again, but the patient's brain was able to maintain its function even as it gradually shrank to less than half its original size. The only reason the doctors could give for this miraculous state of affairs was that—according to Max Muenke, a pediatric brain defect specialist at the National Human Genome Research Institute in Bethesda, Maryland—"different parts of the brain take up functions that would normally be done by the part that is pushed aside."[66]

Cases like that tend to falsify what we think we know about the brain's relationship to consciousness, at least insofar as classical brain anatomy is concerned. But before we can address the anomalies we have to look at what we already know about the neurons and the role they play in consciousness.

The neuron is the basic cell structure we find in the brain. In fact, there are *17 billion neurons* in the human cerebral cortex alone, and as many as 30 billion neurons in the brain overall.

What Is a Neuron?

A neuron is a cell. A nerve cell. It consists of a nucleus (with its allotment of DNA), cytoplasm (where you will find the ribosomes, which contain RNA, as well as the mitochondria, which supply the energy by oxidizing glucose), the cell membrane, the all-important dendrites, and the axon (see Figure 13). Dendrites ("little fingers") are the wires that extend outward from the neuron and conduct most of that cell's business by receiving information and sending it down the body of the neuron along its axon, at which point it communicates with the dendrites of other neurons across the synapse—a word that means "to clasp," indicating the space between one neuron and the next, across which the information is transmitted by means of an electrochemical impulse.

The "chemical" in "electrochemical" refers to the neurotransmitters, which are called that because they are the substances that the neurons use to communicate with each other. There are about fifty different neurotransmitters in the human neuron, located in sacs, or vesicles, in the axon. The neurotransmitters are released by the axon to receptors on the dendrites of another neuron across the synapse. The way this is done, and the employment of electrical charges to do so, is nothing short of amazing.

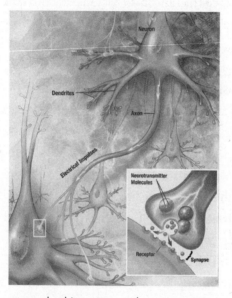

Figure 13. The neuron, dendrites, axons, and neurotransmitters.

Some of the more well-known neurotransmitters include adrenalin, dopamine, serotonin, and the endorphins. They are responsible for everything from mood changes to energy to motor function, sleep and wakefulness, body temperature, etc. They can be excitatory (stimulating a response) or inhibitory (suppressing a response). We can trace many of these neurotransmitters to genetic precursors. For instance, dopamine is synthesized from the amino acid tyrosine, and serotonin's precursor is the amino acid tryptophan. Other neurotransmitters include the "unconventional" endocannabinoids that bind to cannabinoid receptors on the neurons, which among other things respond to the use of cannabis and THC, a pharmacological substance that affects consciousness directly. The command to release the neurotransmitters comes from messages received at the dendrites and passed through the body of the cell—called the *soma*—and down the axon, where the neurotransmitters are found. This is both a chemical and an electrical process, involving the interplay of positively and negatively charged ion particles both within and without the cell's membrane acting as a kind of pump.

The axon itself is like a computer cable. In the white areas of the brain they are seen insulated by sheaths of myelin. This fatty material keeps the signal of one axon from being contaminated by the signal of another and thus screwing up the data; it also helps to increase the speed of the signal being propagated. (Axons without the myelin sheath are known to operate at slower transmission speeds.) Deficiencies in the myelin sheath are responsible for many neurodegenerative diseases, such as multiple sclerosis. Certain pesticides are known to cause demyelination as well, such as the organophosphates.

These neural pathways are critical for any kind of sensory processing. If the neurons stop firing there is no functioning brain, and thus no discernible consciousness. One simply cannot appreciate the vast number of interconnections between these 17 billion cortical neurons (often described as more than there are stars in the Milky Way galaxy), and how their computing power can shift from one area of the brain to another if circumstances demand it. It is as if the sensory inputs keep coming in, regardless of the condition of any single neuron or group of neurons. If part of the brain is damaged, often another part will pick up the slack so that the sensory data keeps getting processed. There are conditions, however, from which it is virtually impossible for the brain to recover. Myelin deterioration, which leads to disorders such as Alzheimer's disease, is among those.

Thus, the physical brain and its nervous system can be affected by conditions in the environment such as trauma, pollutants, neurodegeneration, drugs, and even physical exercise. Does that mean that consciousness is likewise affected? If so, is consciousness nothing more than the program being run by our computer-like brains? When our brains are damaged (as computers are), does the program stop running?

In fact, is consciousness nothing more than computation, or is computation only a platform for consciousness?

Again, we come to that thorny problem of defining and identifying consciousness. How can we say that damage to the brain affects consciousness if we don't know what consciousness *is*? We are meant to understand "consciousness" in an instinctive way, but that's not very scientific. We make a lot of assumptions about consciousness that have no real foundation beyond our feelings about it (it's like that old saw: "I can't define pornography but I know it when I see it"). Let's see if we can come any closer to a resolution of this problem.

Recent studies of the way in which neurons function have suggested that the human brain is not a computer at all. In fact, it is the neuron itself that functions like a computer and that, indeed, is more sophisticated and has greater processing power than any computer available today. The implication of that discovery is that the human cerebral cortex contains 17 billion *computers*. And all of those computers are communicating with each other in a kind of massively parallel way. To make matters even more startling, it has been suggested that the cortex is not merely a collection of 17 billion computers but is "a neural network of neural networks."[67] The explanation is beyond the scope of this study, but suffice to say that some neurons compute nonlinear functions as a self-contained multi-layered neural network. In other words: a single neuron is a multilayered network computing nonlinear functions. Multiply that by billions of neurons all computing at the same time . . . and those billions of neurons are connected to other neurons . . . and, well, my abacus doesn't have enough beads.

This all goes to say that while we believe we understand something about the brain and how it works, we are finding out every day that we really don't. We are still a long way from discovering how consciousness works, but at least we have a platform on which to stand: billions of tiny, super-powerful computers locked away in our heads, running systems, programs, maintenance software, and

God-knows-what-else. But why? What is the purpose? And how did this come to be?

The neurons come from somewhere and are organized incredibly well. That "somewhere" is the DNA molecule, which produces everything that we have in the human body. Does that mean that DNA is the source of consciousness? If we dig deeper we might find the answer or, at the very least, more questions. There are some researchers who claim that DNA is, indeed, the source of consciousness because DNA is the source of all life on Earth. If DNA issued the commands to build the neural networks of the brain and spine, and if the neurotransmitters themselves have genetic precursors, then DNA already knew exactly how to do that: how to build an incredibly complex system of 17 billion neurons in the cerebral cortex. That knowledge must be encoded in the microscopic genome as a series of commands. If it is—and there is no conceivable way it isn't—then DNA must be where consciousness comes from. If we can just understand DNA we can understand ourselves, and if we understand ourselves we can begin to understand the Others.

THE EVOLUTION OF CONSCIOUSNESS

We are not Aristotelian–not brains but fields–consciousness. The inside and the outside must speak, the guts and the blood and the skin.

– Jack Parsons, UFO experiencer, rocket scientist, and founder of the Jet Propulsion Laboratory, in a letter dated January 27, 1950.

THERE ARE ESSENTIALLY TWO SCHOOLS OF THOUGHT where consciousness is concerned. One school insists that consciousness is an "emergent property of the brain." In other words, as brains evolved, so did consciousness; therefore, before there was a physical brain there was no conscious being. This school believes that consciousness is computation, and that better and faster computers will soon become conscious as they reach a technological threshold known as the "Singularity." At that time, it will be impossible to distinguish a machine from a human being. This is the school to which many (if not most) scientists belong, including the very influential scientist and philosopher Ray Kurzweil.

The other school of thought rejects the "emergent property" argument, because it rejects the "consciousness as computation" claim. Since consciousness has never been defined by the first school, so the argument goes, to claim that a machine will one day be "conscious" is mere speculation. It is not based on science, since no actual threshold between computation and consciousness has ever been identified or quantified. This school believes that consciousness exists outside of the classical world and comes from the quantum world, at what is known as the Planck scale, which is where the physics of the classical world cease to apply and the quantum world begins, or at 1.6×10^{-35} meters. This is the school best represented by the theory known as Orch OR (to be discussed later).

To help us understand the relevance and implications of these points of view, we should start with this basic question: when did consciousness first appear on Earth?

In 2012 and 2013 a number of papers were published in peer-reviewed journals reporting evidence suggesting that what we call consciousness is the product of an evolutionary process that began during the Cambrian Period, i.e., more than 500 million years ago. This theory is based on examination of vertebrate fossils that show the rudimentary beginnings of a spinal cord and brain at that time and is focused on the lamprey: a jawless fish resembling an eel in appearance but having a toothed sucker for a mouth (a lot like those sandworms in *Dune* but much smaller!). A modern descendant of the Cambrian lamprey exists to this day and is often considered more of a pest than an ancient artifact, although some cultures find lampreys to be a good food source. (Just ask the Queen of England, for whom lamprey pie is a delicacy.) The idea that the lamprey's ancient ancestor is the locus for the earliest known appearance of consciousness is based on the conviction that some kind of brain and spinal column is a prerequisite. In other words, if consciousness has a physical basis, it must be the brain and the nervous system. Since the earliest known appearance of a brain and spine is found in the fossil record of the Cambrian Period, then consciousness must have started at that time. The corollary to this is, of course, that there was no consciousness on the planet until the humble lamprey appeared.

This is still a "hypothesized" origin of vertebrate consciousness and not yet a definitive finding, but the underlying assumptions are important for this study. In a recent paper on the origins of consciousness by Feinberg and Mallatt[68] we discover that those authors are in agreement with the much earlier assessment by Francis Crick and Christof Koch[69] that the visual system offers the most fertile ground for research in the origins of consciousness and how consciousness is constructed by the brain.

This is because the mechanism of the eye and the response by the neurons is so complex yet immediate. For instance, the eye perceives a circle as an ellipse, but the neurons in the brain will go through a kind of library of possible shapes and correct the image recorded by the eye, and will do that in nanoseconds. It will then track the movement of that circle through space as, for example, you walk over to a plate, pick it up, wash it in the sink, and set it on the rack to dry. That circle will move through various angles of perception as it is picked up, turned over, placed on its side, etc., but will never lose its circular appearance. All of that is due to the firing of millions of neurons that (a) recognized the geometric shape, (b) selected "plate" as the identity or purpose of that particular shape, and (c) followed the shape through your field of vision. This is an oversimplification of what transpires, of course, but you get the idea. While the eye is identifying the shape as a circle and as a plate, your motor skills are being enlisted to pick up and wash the plate. This too relies on the eyes, which have depth of field and stereoptical capability and can direct your hands to the appropriate spot in space where the plate is located and then manage your handling of the plate. Your brain will also identify the color of the plate, its weight, its size relative to you and your environment, and the sound it makes when you drop it. Neurologists understand these mechanisms in their "macro" incarnation but not so much at the "micro" or neuronal level. The latter is where a great deal of research is still taking place.

And the nagging question behind all of this complex computation remains: why did you decide to wash the plate in the first place? What led to you make that decision? What is your particular history, your memories of washing foreign objects, the cultural implications of a dirty plate versus a clean plate, and the values associated with "dirty" and "clean"? And so on.

At the same time, researchers based in Edinburgh agree that the origin of consciousness occurred in the Cambrian Period, but they point to a "genetic accident" at that time that increased the number of "brain genes."[70] This accident occurred, they claim, to a *Pikaia*: a creature similar to a hagfish that had developed an early form of the spinal column.[71] The additional "brain genes" resulted not only in an increase in brain size and intelligence but also an increase in the possibility of mental disease. To summarize drastically: with a larger and more complex brain and nervous system, the chances for something to go wrong increase as well. These same researchers suggest that disorders as varied as autism and schizophrenia may be traced back down the evolutionary path to that initial "genetic accident."

They trace the path of genes specific to "post synaptic density" (PSD) and "MAGUK[72] associated signaling complexes" (MASC)—which regulate protein synthesis and structural plasticity and underlie "learning and memory"—from the invertebrates to the vertebrates, and from the vertebrates to cranial specific development: i.e., from creatures possessing only a spinal column and perhaps a rudimentary brain to a fully functional brain and spinal column. Protein synthesis is a prerequisite for genetic development and complexity. By tracing those elements, one can reasonably assume that consciousness followed the same historical path as brains and nervous systems, which became correspondingly more complex and articulated. Again, the importance of a brain and spinal column—the basic anatomical components of the central nervous system—are considered markers for the evolution of consciousness.

There are, of course, a number of assumptions built into this research. Foremost among them is the belief that consciousness requires a physical substrate. To a materialist, there is no room for doubt on this issue; it is obvious. When the brain dies, the "person" dies. One can keep the body alive using artificial means, but brain death is considered death because—it is assumed—once the brain stops working there is nowhere for "consciousness" to go. It certainly isn't to be found in that particular brain or that particular body any longer.[73]

This assumption would seem to be buttressed by another unassailable fact: the more complex the brain and nervous system, the more advanced or sophisticated the "consciousness" it displays. That would imply that consciousness is dependent upon the physicality of the body. If the lamprey is conscious with only a very basic spinal column and brain, it is not as conscious—certainly not conscious in any way we could appreciate—as a cat or a dog, much less a human being. Thus, the more advanced the *body*, the more advanced the *consciousness*; consciousness is an artifact of the body, which is an artifact of the genes, which are themselves an artifact of chemistry.

Again, this takes for granted that human consciousness is superior to all other forms and types of consciousness on the earth. One of the factors involved in this estimation is the fact that humans possess—besides the brain, spinal column, and nervous system—four limbs, two on each side of the torso, and a jaw. The jaw is an indicator of the type of vertebrate scientists consider essential to the development of human-type consciousness. The symmetry of the limbs on either side of the torso is another factor, indicating as it does a connection with the bicameral brain: the two hemispheres, right and left, that control the left and right sides of the body, respectively, as well as different types of consciousness (typically, the left brain is said to "control" rational and logical thought, whereas the right brain "controls" imagination and creativity). This enabled creatures to reach out into the environment and collect information—through the limbs, hands, and feet in addition to the eyes, ears, nose, mouth, and skin—and to develop a binary approach to that information: right/left; up/down; forward/backward; in front/behind; and, possibly, past and future. Creatures with four limbs could travel great distances on the ground and collect valuable information along the way, extending their range and making new discoveries. In turn, learning and memory increased; with such an enormous storehouse of information at their disposal—the earth—the vertebrates began to consume data as well as other resources. And, eventually, the development of the opposable thumb added even more capability and the ability to make tools.

To say that the limbs and senses are extensions of the brain and nervous system would be to put the cart before the horse. It is just as logical to assume that the brain and nervous system are the result of the development of the limbs and senses. In reality, the two systems—the anatomical and biological on one hand and the nervous system on the other—developed together, each reinforcing the other, giving us enormous power over the world we live in. This means that certain capabilities we may have had eons ago were selected out during the evolutionary process, as they did not contribute to our survival or perhaps made us weaker. As we survived, we grew stronger not only physically but also intellectually and creatively: both sides of the brain cooperated to give human beings tremendous range.

It took five hundred million years, but those simple vertebrates and pre-vertebrates of the Cambrian Period eventually yielded up a creature that could communicate over vast distances in an instant and travel to the Moon and beyond.

At the same time, though, science is discovering that the identical systems that gave rise to our enormously complex brain also gave it some serious vulnerabilities, as mentioned previously, among them mental disorders such as Alzheimer's disease. The ancient immune system of antimicrobial peptides that protects the brain was found to turn on it with age: it creates a molecule that forms the type of plaque that results in dementia. It is a process that is still not understood but has implications for any study of neuroscience. With the growing complexity of the cerebrospinal system came greater possibilities for things to go very wrong. What does that mean for consciousness itself? Like asking, "Is the thought of a unicorn a real thought?," we may ask if a consciousness in a brain affected by Alzheimer's disease (or any one of hundreds of other mental disorders, including autism and schizophrenia) is real consciousness. Is there a baseline consciousness against which all other forms may be measured?

Is it possible, therefore, that some of those genetic characteristics that we once had millions of years ago and lost because they were not essential to survival were discarded a little too hastily? Is it possible that what made us vulnerable then—in an environment of savage beasts, an erratic and threatening climate, and hostile neighbors—might be useful to us today? We are part of the history of the world and its trajectory. Who we are owes as much to where and when we live as to our genetic makeup. Now that the world itself is changing in so many ways (and now that we ourselves are changing it) did we leave something valuable behind?

There is some evidence that we did, but that the genetic characteristic we "lost" has been hiding in plain sight all along: in our ancient literature, if nowhere else. What we call "consciousness" today might actually be a somewhat scaled down (or, at least, considerably altered) version of what we once possessed.

We trace our consciousness to that period in ancient prehistory when the lamprey (or the hagfish) developed vertebrae . . . and we stop there. We make the assumption that consciousness did not arise until a working nervous system was available. But what if consciousness was already present in the genes themselves, which contributed to the developing complexity of the nervous system, driving it on to greater and greater achievements? Why stop at the nervous system? Why even stop at the genes?

Does having a primitive radio receiver that only gets one station mean that there is only one station? In other words, did consciousness evolve as an artifact of biological or genetic complexity, or did consciousness become more *accessible* the greater the degree of biological or genetic complexity? To use the admittedly tired old computer analogy: do we understand the computer operator by taking apart the machine itself? The CPU is intelligent; the RAM and ROM answer the need for memory. There are input and output devices that act as sensors and limbs in the environment. But until an operator turns it on and issues instructions, the device is inert. It is an inorganic instrument that responds only to an organic operator somewhere down the line.

It was precisely this sense of the human being as a kind of machine—a creation of some other, superior intelligence—that inspired Zecharia Sitchin to write his famous series of books in which he claimed that humans were created by aliens to mine gold on Earth. And it may be that same suspicion that has contributed to the tales of the Grays being controlled or commanded by taller beings that some call the Nordics. So, which one of us—human beings, Grays, Nordics, or any of the dozens of other proposed alien races—are machines? While we still have difficulty defining consciousness, we slowly come to the realization that we face the same difficulty when it comes to defining machines. If DNA was seeded on Earth, was that done by a machine? And was the purpose to create new iterations of self-generating automata?

In other words, is the human race a "sekret machine"?

The Quantum Controversy

For many of us, consciousness is the container for emotions as well as intellect. It is also how we interact with notions of the divine, of life after death, and with paranormal experiences for which there is no physical proof or tangible evidence. Evolving from the earliest pre-vertebrates, we arrived where we are now with very strong ideas about invisible things: gods, spirits, demons, jinn. These ideas have become confabulated with accounts of UFOs and extraterrestrial biological entities (EBEs). UFOs are invisible most of the time; that is, we don't see them. But they are visible some of the time and are seen, experienced, tracked, and investigated. Same with the EBEs or "Visitors." Our brains are superbly capable of ordering the inputs coming in through our senses and making sense of them so that we can react appropriately. But when it comes to the Phenomenon, we are left on very shaky ground as our neurons struggle to make sense of what we are experiencing. For such a sophisticated computational system as the neuron to fail us in this regard seems somewhat suspicious.

Most of human experience proceeds along a historical continuum. We see things that fit into a pattern of what we know. When we see a bird, for instance, it fits within the context of all the birds we have ever seen, whether in real life or in books, other media, etc. When we see a person we have never seen before, the brain identifies it as human, and also identifies its gender, its age (perhaps relative to our own), and so forth. There is a context to everything we experience. The neurons are prepared for this. New experiences are matched against memories of previous sensations and the model is cleaned up and presented to us in fractions of a second. Then we go along our way, oblivious of the intense activity that is taking place behind our eyes.

With the Phenomenon, however, our neurons work frantically to find something that the experience is "like." Is it *like* a specific color, or shape, or movement? The Phenomenon is not contextual; it is not historical in the sense that it is not an active player in human history the same way that climate, economy, violence, technology, politics, race, etc., are identifiable and everyday forces at work in our world. The Phenomenon is completely unpredictable: it shows up without warning, acts in accordance with physical laws that are not consistent with what we know of reality, and according to norms of behavior that are foreign to every culture on the planet, and then disappears without warning.

The Phenomenon involves both sightings of actual machinery—the UFO/UAP themselves—as well as individual beings, the so-called "aliens" or Visitors. Both the machines and the beings behave in ways that defy the laws of physics as we understand them, and at times they even seem to flaunt their capabilities in front of us, as when they follow commercial and military aircraft as if daring the pilots to follow them or shoot them down.

But what if they are not violating laws of physics at all? What if they have somehow mastered a quantum effect that we have not discovered? They do seem able to communicate with us in a way that defies normal avenues of discourse. What if that, too, is a quantum effect?

What if we, human beings, are being teased into discovering the same effects?

▼　　▼　　▼

Many scientists reject—or openly ridicule—the notion of "quantum consciousness." They believe that the "spooky" nature of quantum physics (including especially non-locality, superposition, and the like) is sufficiently misunderstood by the layperson to be associated with other "spooky" phenomena, such as mental telepathy, ghosts, and UFOs. Laypersons who wish to use quantum effects to explain ESP or remote viewing or other paranormal phenomena are considered ignorant, gullible, or just, well, irritating.

It is true that many enthusiasts of the idea of quantum consciousness do not have the training in classical physics—much less quantum physics—to be able to explain and defend their position on the paranormal. There is, however, a level of arrogance in the abrupt dismissal of their claims by scientists that only creates more conflict and increases the level of misunderstanding and ignorance.

Critics of quantum consciousness theories will point to some obvious examples as containing misunderstandings of how physics actually works and how physicists measure, study, and test quantum effects.

There is tremendous resistance on the part of physicists where extrapolating from individual test results—such as in the famous "double slit" experiment—to imagining their implications for real-world applications is concerned. Their argument runs that quantum effects occur only at the sub-atomic, quantum level (the Planck scale) and while the laws of quantum physics seem weird and spooky they are not applicable on the macro level, where classical physics applies. In other words, once we are dealing with atoms and molecules, the quantum effects are no longer observed. At the macro level we deal with the kind of classical physics with which we are accustomed in the everyday world: a world where travel or communication faster than the speed of light is impossible, where an object cannot be in more than one place at the same time, and where the cat in the box with the plutonium is always dead.

This was especially true of the insistence by some scientists that quantum effects occur in the human brain and might be responsible for consciousness. "Nonsense," was the usual response. "The brain is warm and wet and noisy; quantum effects can only occur in extremely cold and dry conditions." When Sir Roger Penrose and Stuart Hameroff proposed their "Orch OR" theory of quantum consciousness in the late 1990s, they were shot down because of the assumption that quantum mechanics requires cold and dry conditions.

And that was pretty much the state of affairs until 2014. At that time, new research began appearing in peer-reviewed journals that reported quantum effects in photosynthesis as well as in the brains of migratory birds. Penrose and Hameroff felt vindicated, and began defending their discoveries with greater enthusiasm.

We won't go into all the proofs and arguments for the various theories—for which we assume the reader will be grateful—but we will look at why quantum physics provides what may be the best approach toward a theory of consciousness, both from the purely philosophical point of view and that of physics itself. It's possible that "quantum consciousness" may satisfy the requirements of both the mystic and the rationalist.

Playing Dice with the Universe

Einstein is often quoted as saying that he did not believe that "God plays dice with the universe." This was a remark directed squarely at quantum physics, or quantum mechanics (QM) as it is commonly called. That is because QM at its heart is concerned with probability. For deterministic systems, such as the Newtonian physics with which we are all familiar from high school science classes, the idea that the underpinnings of the physical world are not predictable but are based on "chance" is, well, anathema. Yet QM has been proven many times since Einstein's famous remark, and it is the problem of reconciling QM with Newton and Einstein (especially Einstein) that is making a lot of people crazy. It is as if there are two completely different sets of physical laws governing the universe and they contradict each other in important ways. It is considered counterintuitive, but that may be because our "intuition" is the product of our evolution and of the evolution of our sciences.

Much of what we know about QM comes from popularizing accounts that focus on specific statements or on controversial experiments. Some readers will have heard of Schrödinger's Cat, or the Heisenberg Uncertainty Principle. These examples seem like disconnected pieces of the overall theory, neither of which by itself can bring us any closer to a grasp of what QM is all about.

But, hey, it's worth a shot.

Take for instance the central concept of particles and waves. In QM, everything is made of particles (which is something we can get behind, since we learned about atoms and molecules in school). But everything is also made of waves. Whether something is seen as a particle or a wave depends on the observer, which seems to imply that consciousness plays a role.

This goes back to the "double slit" experiment that has been performed many times over the years and thus proves this essential fact of quantum mechanics. Briefly, when photons—light particles—are shot through a narrow slit in a board (just narrow enough to permit a single photon to pass through at a time) and hit a screen on the other side they create a vertical pattern of dots that resembles the vertical slit they passed through. Okay, that's reasonable and to be expected.

But when you set up two slits, parallel to each other, so that you would expect to see two vertical bands of dots on the screen, something strange happens. You get *multiple* bands of dots. You get an interference pattern. How is that possible?

Imagine throwing a ball through a slit so that it hits the wall, making a mark. Now imagine throwing balls through two slits. You would expect marks on the wall from both balls, and they would be found opposite the corresponding slits. You should be able to draw a straight line from yourself—throwing the ball—through the slit to the wall. But in this case there are multiple marks on the wall in different areas.

That's because the photons (the balls) behave as waves as they pass through the slits, and as waves they create interference patterns with each other (such as the ripple effect you might see when you throw two stones into a lake). Those interference patterns result in many vertical bands of marks on the screen; many more than two slits logically would produce (see Figure 14).

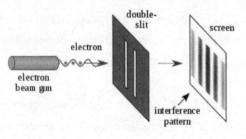

Figure 14.

This confused experimenters, so they placed a sensor at the slits. This way, they could see what a photon did when it left the slit and headed for the screen. They wanted to know how a photon could suddenly transform itself into a wave: a non-particle.

But when they observed the passage of the photon through the slit, it remained a particle and did not become a wave.

Say what?

The act of measuring changed the result of the experiment. The observer actually exerted some kind of influence or force or . . . something . . . over the photon, making it behave differently from the way it would have had it not been observed. In QM, they describe this event as the "collapse of the wave function": i.e., when a wave "becomes" a particle.

This has profound implications for the way we understand the "real world." It also introduced the idea that a photon can be both a particle and a wave (or, perhaps, a third "substance" that partakes of both wave-like and particle-like qualities), depending on who's doing the measuring or observing. It might be more appropriate to say that the photon is neither a particle nor a wave but a disturbance in the quantum field. That leads us to the next principle of QM: uncertainty.

When we studied physics in school, we became familiar with diagrams of atoms. We saw that every atom has a nucleus, and around that nucleus various particles are in orbit like planets around the sun. It's a nice schematic, but it is useful only as a learning tool and does not represent the reality, which is that the nucleus is surrounded by a "cloud of probabilities" and not discrete orbits of electrons.

This reality reflects the realization of quantum physicists that it is not possible to predict the outcome of any QM experiment (or the precise location of any electron) except as a kind of statistical probability. One can locate a particle either in space or in time, but not both.

To add insult to injury, QM is also non-local. This is often explained by saying that when two particles "meet" and then go their separate ways, they are in constant contact with each other even over incredibly vast distances. Sounds romantic, right? An operation performed on one of the particles will instantly affect the other particle, no matter where in the universe or how far away it may be. This is called "quantum entanglement." The problem with this discovery is that it seems to violate the general theory of relativity, which states that nothing can travel faster than the speed of light; in quantum entanglement, the speed of light is not relevant. Information—the state of a given particle at a given moment—is transmitted immediately to the other particle and modifies its behavior appropriately.

If a particle in, say, New York is made to spin clockwise, its partner in Tokyo will immediately begin spinning clockwise, too. There is no delay, no lag time, between the two. This has been tested time and again. The transfer of information from particle "A" to particle "B" takes place at a velocity that obviously exceeds the speed of light, which should not be possible. This "non-locality" is an essential feature of quantum mechanics. There is no "local" and "distant" where QM is concerned. This has led some quantum physicists to insist that what takes place between the two particles is information transfer, and that information is not the same as matter (is neither a particle nor a wave) and thus its "velocity" is not bound by the speed of light.

This is all complicated by the fact that there are multiple particles of matter—mostly electrons, quarks, neutrinos, and muons—and multiple carriers of force such as photons, bosons, and gluons. So just when we thought the Periodic Table of the Elements had us covered, we realize that there are deeper and deeper levels to reality that are composed of smaller and smaller units of matter and force. Moreover, these smaller units do not behave the way larger units do. They operate according to a completely different set of laws, and trying to integrate quantum mechanics with the standard model has so far proved to be impossible.

But these discoveries in quantum mechanics have excited and inspired a generation of mystics, gurus, and con men. Ideas like non-locality, quantum entanglement, and uncertainty seem to create a space for the operation of paranormal abilities and provide a rationale for the belief in ghosts, spirits, and UFOs. After all, goes the theory, with non-locality and entanglement, such things as teleportation, astral projection, and telepathy must be possible.

It's quite a leap from quantum entanglement to astral projection, however. Basically, all you are doing is linking two examples of "spooky action at a distance" and making a false equivalence. It also requires extrapolating from the case of quantum mechanics—which affects subatomic particles only—to the larger world of direct experience.

That said, new research has shown that quantum effects can be observed in the world at large. One case in particular demonstrates quantum effects, and its implications are profound.

That case is photosynthesis. By 2007, a number of articles began appearing in the scientific press that suggested that the process by which sunlight is converted to energy in plants is quantum mechanical in nature. Plants do this via molecular vibrations in plant cells, resulting in energy exchange, which is a process involving negative values in probability distributions, that are unknown to classical physics (which recognizes only positive probability distributions). By 2014, this theory, based on the evidence, had been accepted widely by the scientific community.[74]

This was critically important for several reasons. In the first place, we are now able to say that the effects of quantum processes can be observed in the "real" world: the visible world all around us, and not just at the nano level. In the second place, it demonstrates that quantum effects can occur at biological temperatures (warm and wet) and not only under super-cold and super-dry conditions. This has tremendous implications for the study of consciousness, as we will see.

There is also something oddly compelling about a quantum interpretation of photosynthesis. Since the nineteenth century we have known that light comes to Earth from the Sun (in the form of photons) and then penetrates the cell membranes of plants, enabling them to synthesize food from carbon dioxide and water. This happens when a photon is absorbed by chlorophyll, which then loses an electron and passes it down an electron chain during the process of photosynthesis. In dealing with photons and electrons we are already in the subatomic world of quantum mechanics, but so far classical interpretations have been used to try to explain how photosynthesis works. Unfortunately, efforts to describe the process using classical theory have not been successful.

The idea that our food (and our oxygen, and therefore life itself) depends on the flow of photons to the surface of the earth, where energy transfers take place on the subatomic level, suggests that there may be other mechanisms at work in our world that best can be described by recourse to quantum mechanics. It may in fact be the interplay of quantum forces that creates the web of connectivity between humans and nature, between humans and other humans (at various levels, some of which may be unseen), and between humans and members of other species, both terrestrial and extraterrestrial or nonterrestrial.

Physicists are faced with the dilemma that these two worlds seem to be incompatible, as if reality on the subatomic level is not the same as the reality on the classical, macro level. This incompatibility has contributed to the irritation and frustration many scientists express where claims of a quantum mechanical basis for paranormal states and experiences are concerned. They need the classical physics of Newton and Einstein to be at least an expression of the quantum world of Max Planck, Niels Bohr, Heisenberg, Schrödinger, and Pauli. There cannot be "two thrones in Heaven," to borrow a term from Jewish mysticism. There cannot be two realities. There can't be a world where paranormal phenomena exist and one where they don't, not if we are expected to live in both worlds simultaneously!

The solution might be both easier and more dangerous than we imagine.

Think of a chessboard. It is flat and two-dimensional. The action of chess, however, involves three-dimensional pieces of various designs all moving about on the two-dimensional surface. What we see—what we focus on—are the chess pieces and their interplay with each other in complex patterns. However, without the chessboard, the pieces cannot move. Without the sixty-four squares of the board, the pieces are irrelevant. If we think of the quantum world as the chessboard and the Newtonian world as the chess pieces, we may get closer to an understanding of how these two seemingly separate physical environments actually work together to create reality as we know it. In more modern (and culturally significant) parlance, the chessboard (and hence the quantum world) is the matrix.

The two worlds are not separate from each other: each needs the other to fulfill its function. Taken together they represent the game of chess. Taken separately they can be examined, analyzed, and measured . . . but we learn nothing about the "game" that way. The structure and pattern of the chessboard "bleeds into" the play of the pieces, though, just as quantum mechanics "bleeds into" the problem of photosynthesis.

The rules of chess and the action of playing chess might be analogous to the physics of Einstein. Whereas the pieces themselves are emblematic of Newton's physics, the movement of the pieces and the interplay of the two sides in a game of chess are closer to relativity.

We can take the analogy further, of course, and suggest that the red squares on the board represent particles and the black squares waves; or the red squares represent matter and the black squares dark matter; etc., etc. As with any analogy, you can take it too far. It is only necessary to think of the chess game in its entirety as a way of understanding the role that quantum mechanics plays as a kind of background for classical physics and Einstein's relativity. There is another way of looking at this analogy, however, and it relates to what we discussed in the first section of this book: the genetic code.

You may remember that we pointed out the similarity between the chessboard of 64 squares and the matrix of 64 possible combinations of the four amino acids in groups of three that give us the codons that make up the genetic code. We also noted the similarity between that mathematical pattern and the 64 hexagrams of the Chinese *I Jing* or Book of Changes, and the further development of that pattern into the 256 odu (64 x 4) of the Yoruban *Table of Ifá*.

If the DNA molecule was itself a pathway into the quantum world, it may show us how consciousness, genetics, and quantum mechanics are related on a deep level, and thus how the events in our lives may be like chess pieces played across an existential board. It may suggest to us that the quantum world is the unconscious mind of the world of classical physics.

Fortunately, recent experiments on quantum entanglement in the DNA molecule are suggesting a path to that conclusion.

At the same time that quantum effects were being discovered in the process of photosynthesis, researchers were also seeing quantum entanglement in the DNA molecule itself. Quantum entanglement is that characteristic of QM in which a single wave function describes two separate and distinct (and even distant) objects. No matter how far apart those two objects are—as we have seen above—they weirdly share the same existence. They are "entangled."

In 2010, Elisabeth Rieper and her colleagues at the National University of Singapore theorized that the nucleotide of the DNA molecule is actually a cloud of negatively charged electrons surrounding a positively charged nucleus. The interplay of the negatively charged cloud and the positively charged nucleus creates a kind of oscillation, or *phonon*. In order for the nucleotides to bond and form a base pair, their respective clouds must oscillate in opposite directions. But phonons, you see, are quantum objects, and quantum objects become entangled.

Which means, to make a long and very technical explanation short, that DNA is entangled. According to the theory, it is that entanglement that actually holds the double helix of the DNA molecule together.

While this is not yet "DNA consciousness," it does suggest that there is a quantum aspect to the genetic code. This is important because it is another indication (along with photosynthesis) that quantum effects may be obtained at atmospheric levels rather than the extremely cold temperatures at which quantum effects are usually measured. The implications for building quantum computers, for instance, are considerable.

There is another—even more relevant—conclusion that can be drawn from these theories and experiments, which is that quantum effects could be occurring in the human brain and affecting (or being responsible for) consciousness. If that is true, humanity may be on the verge of reconciling the conflict between two opposing points of view concerning reality: the purely reductionist, deterministic approach typical of what we think of when we think of science and scientists, and the purely "spiritual" or intuitive approach that we associate with the non-scientist or even the anti-scientist. The rational and the creative—like the left brain and the right brain—are two halves of a single reality, and it may be our responsibility as human beings to find a middle way between them.

The Penrose-Hameroff Theory

There is a great deal of discussion in scientific circles where the field of artificial intelligence (AI) is concerned. As we have described several times already, many scientists—perhaps the majority—feel that the brain is nothing more than a very sophisticated machine, a computer, and that as computer technology becomes more advanced, computers will themselves develop consciousness. In other words, consciousness is nothing more than an "emergent property of the brain." As the brain evolved into a more complex instrument, so the theory goes, consciousness developed as a kind of "secretion" of the brain. In this view, animals may be conscious to an extent but plants are not. Plants have simple systems that react to their environment—to sunlight, to water, to temperature—but it is not consciousness the way we understand it. After all, a compass magnet will point north, but that does not mean the magnet is choosing to point north or that there is consciousness in the magnet. A plant bending toward the sun does not necessarily imply a conscious reaction or "decision." In the first place, it cannot pick up and move from one area to another, and it is thought that consciousness depends—at least in part—on the ability to interact actively with one's environment. This ability requires the development of senses that plants do not have, such as sight and smell and hearing. (For instance, we say of persons who are in a coma that they are in a "vegetative" state.)

As we have seen above, our sensory systems are an integral part of our central nervous system (CNS). So it would seem that the development of a nervous system—spinal cord and brain—would be necessary for consciousness, as those who study the evolution of the brain insist. From this perspective the development of true artificial intelligence should require that next-level computer systems be equipped with additional sensors to mimic, if not duplicate, the human CNS. If a computer simply sits there on a desk or is hard-mounted in a laboratory, it is doubtful whether that computer would ever really approach human consciousness; it may instead remain in a "vegetative state." However, the proponents of "strong AI" are confident that, with greater and greater algorithms and faster processing power, computers would acquire consciousness—they are, after all, based on the same physics as the human brain— and that eventually human and machine would become integrated to the point that we would not be able to distinguish one from the other. This is the point of view of inventor and philosopher Ray Kurzweil, for instance, as represented in a number of books and papers he has published over the years.

The opposing view is that computers will never attain human-type consciousness because the science of artificial intelligence is missing a few essential pieces of the problem. This view is the one developed by Sir Roger Penrose and Stuart Hameroff. Penrose is a famous mathematician and physicist, and Hameroff is an anesthesiologist and professor emeritus of anesthesiology and psychology. Together they argue that the brain is not a deterministic machine that operates algorithmically, like a computer; that the types of processes of which the brain is capable are not limited to the kinds of programs—which are based on algorithms— used to program computers. They insist that a way must be found to unite quantum mechanics with classical theory in order to explain consciousness.

There is some elegance to this recommendation, actually. Both classical physics and quantum mechanics are discoveries of consciousness, of conscious beings who worked at understanding how the world operates. The laws of physics can be considered inventions or projections of the human mind onto experience, ways of organizing information, but they do not make any statement about the reality of the laws apart from their usefulness to human beings. In other words, the laws don't actually "exist" in any concrete way. They are the product of a train of thought that has been going on in the minds of thinkers for millennia.

So since both classical physics and quantum mechanics are "true" and both are products of human consciousness, it stands to reason that one needs both systems to explain the consciousness they came from.

We already know a great deal about the physical brain and its various systems; now we need to find out if quantum mechanics can fill in the gap in our knowledge of how the brain works. Perhaps then we will understand something more of consciousness. That was the goal of Penrose and Hameroff.

The physical substrate for what Penrose and Hameroff call "orchestrated objective reduction" (Orch OR) is the microtubule.

The microtubule (MT) is present in the cytoskeletons of everything from single-celled organisms to neurons in the brain. The MT is hollow, about 25 nm (nanometers) in diameter.[75] It is composed of hexagon-shaped lattices of proteins (as in Figure 15), and is essential for many cell functions such as movement and cell division. In the neuron its function is in extending the dendrites and axons and forming synaptic connections.

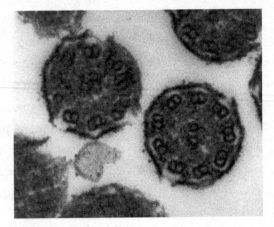

Figure 15.

For quite some time, the importance of the MT was not understood. One of its recently discovered features is the strange coupling of tubulins (the protein that is the main element of the microtubule) within the MT in dynamic switching patterns that suggest the "cellular automata" of the "Game of Life" developed by John Horton Conway in 1970. As more MT data was analyzed it became clear that microtubules were responsible for much more than the structure of the neurons; they were also involved in electrochemical processes including signaling and communication. Suddenly, the MT became the focus of greater attention due to the possibilities that it offers as a medium for consciousness.

For Penrose and Hameroff, the MT might be the Holy Grail that provides the bridge between classical physics with quantum mechanics, as it shows that the human brain demonstrates quantum effects. The implication is that such features of QM as non-locality, entanglement, and super-position may be foundational to consciousness and provide the missing link between a purely mechanistic and deterministic view of consciousness as an "emergent property of the brain"—a result of its computations and nothing more—and the more "spiritual" concept of consciousness as something that exists (and pre-exists) outside the brain. In fact, the very architecture of the brain's neurons may point to the brain as a device for quantum communication:

> Orch OR suggests that there is a connection between the brain's biomolecular processes and the basic structure of the universe.[76]

"Orch OR" is described as:

> a theory which proposes that consciousness consists of a sequence of discrete events, each being a moment of 'objective reduction' (OR) of a quantum state . . . where it is taken that these quantum states exist as parts of a quantum computation carried on primarily in neuronal microtubules. Such OR events would have to be 'orchestrated' in an appropriate way (Orch OR), for genuine consciousness to arise. OR itself is taken

to be ubiquitous in physical actions, representing the 'bridge' between the quantum and classical worlds. . . . In our own brains, the OR process that evokes consciousness, would be actions that connect brain biology (quantum computations in microtubules) with the fine scale structure of space-time geometry, the most basic level of the universe.[77]

If consciousness results, as Penrose and Hameroff insist, from "discrete physical events" that "have always existed in the universe as non-cognitive, proto-conscious events," then "biology evolved a mechanism to orchestrate such events and to couple them to neuronal activity, resulting in meaningful, cognitive, conscious moments and thence also to causal control of behavior." Thus, according to their Orch OR theory, "these conscious events are terminations of quantum computations in brain microtubules . . . and having experiential qualities. In this view consciousness is an intrinsic feature of the action of the universe."[78]

We might say, then, that consciousness is an emergent property of the universe itself. If that is so, then it is reasonable to propose that beings from other parts of the universe also exhibit consciousness if the only criterion is access to "quantum computations in brain microtubules." If we interpret "brain" in the most liberal sense of a central nervous system, then an "alien" brain might exhibit consciousness but have access to different *qualia*—different subjective experiences, and different interpretations of the same experiences—and therefore interpret the world differently.

They would, however, be able to communicate—via the microtubule structure, or some analogous structure—with other conscious beings, if consciousness is really reducible to the types of discrete physical events identified by Penrose and Hameroff. The quality of that communication would be affected by the differences between human brain and neuronal structures and those of the alien or Other species, but it would nonetheless take place. Moreover, if communication between a human being and an alien occurred at the level of the microtubules and thereby involved quantum entanglement, it becomes possible to suggest that the claims of alien abductees that they are always in contact with their abductors (they can be located no matter where they are on Earth, as if they were "tagged") may be based on this type of quantum mechanical phenomenon.

While this discussion may seem impossibly arcane to those not involved with quantum mechanics on a regular basis (which is most of us) there are some takeaways that we can apply to the Sekret Machine project. We know that QM is being explored as the basis for a new generation of computers; it is also being explored as a means to achieve teleportation (so far of individual particles, not entire organisms). The possibility of real-world applications of such quantum effects as entanglement and non-locality is enormous.

Of greater interest to our project, though, is the idea put forward by Hameroff that the universe rises out of consciousness, that consciousness preceded life and "drove its origin and evolution."[79] This is not a new idea, as one can find it articulated at length in Asian scriptures. What is different is the scientific approach to the idea, which for Hameroff means that the evolutionary process is not the "selfish gene" concept of Darwinian philosophers such as Richard Dawkins—a concept that was amplified in this book earlier, as the desire of genes to live forever at the expense of their hosts (us)—but a "pleasure principle" similar to that conceived by Sigmund Freud, and based not on psychology but on the observation that reproduction—either of mammals, or of proteins—comes about not as the result of some ill-defined mechanism of gene immortality and Darwinian survival but as a consequence of pleasure: i.e., because it feels good. To Hameroff, there is a feedback loop that is basic to the universe and hence to consciousness and this is a reinforcement (reward/punish) of certain behaviors at the expense of others.

This seems reasonable if we consider that the gene cannot know the future, and therefore cannot project what reproductive strategies will best serve its long-term goal of immortality, but it can know immediate reward or punishment, pleasure or pain, and function accordingly.

If this feedback loop—which is obviously part of human nature—is universal, as Hameroff implies that it is, then it may undergo different manifestations on the macro level depending on the species involved. In other words, for human beings the pleasure principle is clear and is responsible for sexual reproduction and the propagation of the species. But what if this is not so clear to another, nonhuman, species? What if the evolutionary process on other planets favored a more rational, logical, and pragmatic approach to life, perhaps due to environmental factors that did not encourage the pursuit of pleasure? Just as some animals on Earth have certain reproductive cycles and do not mate during the year except at those specific times, it is possible that an alien species would behave in a similar way, and that human reproductive practices would mystify them. They would try to see how the human reproductive cycle is different from those of, say, cattle.

The experience of abductees with the alien curiosity concerning sexuality might be linked to the famous cattle mutilations that have taken place over the years and which are often linked to UFO sightings. If the alien species is non-organic—an android or robot of some type—this makes the possibility of intellectual curiosity even greater. After all, we humans used to rob graves and dig up corpses of our fellow humans to find out how the body worked. How much more curious about those inner workings would an alien species be?

Would the aliens try to locate the pleasure centers in the human brain and attempt to discover how pleasure worked, and how it was responsible for human reproduction? Would they have a different concept of how consciousness works, one that takes into account neither Hameroff's "quantum pleasure principle" nor Dawkins' "selfish gene," but some other model of which we have no idea? We may have erred in assuming that the alien interest was in our reproductive systems per se, and not in our motivations for behaving sexually.

If Penrose and Hameroff are correct in their claim that the action of microtubules in the human brain are responsible for consciousness—for intrinsically connecting the brain to what they called the "fine-scale structure of space-time geometry"—then how integrated to that fine-scale structure would an alien nervous system be if it did not evolve the same way, with the same (or similar) neural structures, including the microtubules? Would that alien brain be as conscious as a human's? In the same way? Or would a different nervous system actually produce a different form of consciousness altogether?

Since the alien is able to communicate with its abductee—in a communication stream that is largely one-way, from alien to human—we must assume a certain degree of similarity between alien consciousness and human consciousness without being aware of the essential features or vulnerabilities of the alien version.

Perhaps chaos theory can provide another perspective.

UTTER CHAOS

It seems to me that a complete psychology would ask also, for example, why it is that human beings do physics, mathematics, and art *at all*; how are consciousness, mind, and thought related to these activities, and how does it all fit together? From this point of view it would seem to follow that if there is to be a "theory of everything," that physicists are so fond of claiming as *their* territory, psychology will likely have to provide it.

– Larry R. Vandervert[80]

WE HAVE INSISTED THAT A MULTIDISCIPLINARY approach is essential if one wants to understand and explain the Phenomenon. We would be amiss, then, if we did not include other perspectives on consciousness, since it is consciousness itself that is under scrutiny when it comes to eyewitness accounts of UFO/UAP Phenomena and especially alien abductee experiences.

While the theories presented in the previous chapter—especially the Orch OR theory of Penrose and Hameroff—are focused on the physical, neurobiological substrate of consciousness and are invaluable for that reason, here we are going to approach the problem from the lens of a novel theory of psychology that includes physics and mathematics (as does Orch OR and indeed any form of quantum mechanics) but strives to offer a more holistic answer to the problem.

This is chaos theory.

Basically, chaos theory states that beneath the seemingly random or chaotic features of a system there is an underlying pattern including feedback loops, self-organization, and repetition, and that these represent an initial condition upon which the seemingly chaotic features depend. One often overused metaphor for chaos theory is the "butterfly effect": that a butterfly flapping its wings in China can affect weather conditions in Texas; i.e., that a small change in a deterministic and nonlinear system can have a serious impact further down the line as that seemingly minor deviation snowballs.

In the previous chapter we looked at the evolution of consciousness as something that could be measured as an expression of the evolution of the nervous system: a process that began during the Cambrian Period, 500 million years ago. We also looked at the Orch OR model, which claims that consciousness has been here all along and maybe pre-existed the universe.

In this chapter, we will see if chaos theory can lead us into some other interesting pathways.

▼　　　▼　　　▼

In the 1990s, psychologist and neuroscientist Larry Vandervert published a number of papers on chaos theory,

of hallucination. If it is a hallucination, it is one whose details are strangely consistent from witness to witness. The psychologists ascribe that similarity to neurobiological causes. In the modern Western world there is a sharp division between "real" events and "imaginary" events, the "imaginary" events being those that *you* see or experience but *I* don't. In other cultures, however, this division is not so clear, especially in those cultures where self-identity is created or maintained through processes of socialization.

But Vandervert makes an important distinction between "consciousness" and "mind." In his view, animals are conscious but "since they don't have culture they likewise do not have appreciable 'minds.'"[83] This same equation could be applied to an analysis of alien consciousness. Is it possible for us to ascertain whether or not the alien has a culture, which therefore can be said to have contributed to the alien "mind"? And how does this relate to the main thesis of artificial intelligence, which is that machines will eventually become conscious? Where does "mind" fit in?

According to Vandervert's theory of neurological positivism (NP), the mind has two forms. The first is the one with which we are familiar, which is the circuitry of the brain and the nervous system. The second manifests as culturally shareable mental models. Mind, according to Vandervert, began with culture and has evolved—and continues to evolve—as humans evolve. Thus, while the "real world" of objects has not changed, the mind has evolved to create various models of the real world, such as mathematics and music, which become shared cultural artifacts. This is all taking place against the backdrop of the very same space-time geometry referred to in the Orch OR model. In other words, as we create more and more complex models of the world, reflective of our increasingly complex nervous systems and the steps we take to change, improve, or alter those nervous systems through machinery, meditation, psychoactive chemicals, etc., we are affecting/effecting the world we are creating. With every new iteration we are making new models. These "represent the actual relationships among world, brain, and mind"[84] as depicted in the holonomic models proposed by Pribram (and by David Bohm and others).

To illustrate what he means, Vandervert uses the example of the phantom limb.

In this case, there have been people who were born missing limbs yet they are aware of the limb, since the brain perceives the body as intact and the limb in its proper place in space and time. We know that this has happened to those who have lost limbs due to an accident or illness: the limb still is felt to be there and to experience sensations. But why would this occur in individuals who never had the limb to begin with? A researcher named Melzack proposes that this happens because the brain continuously generates a pattern of impulses that represents "a pure space-time template for the body."[85] Vandervert calls the phantom limb phenomenon "a (the) vital clue as to the origin and nature of consciousness we all experience."[86] In other words, the brain (or the mind) continuously creates an experience of the body as a coherent entity, with dimension and movement in space and time, in order to navigate through the world of sensation. The existence of the "phantom limb" is the clue that the brain is functioning this way, with a *model* of the body. The body in the brain. It is a key, as Vandervert says, to the nature of consciousness.

Any portion or piece of a hologram contains the entire hologram; that is pretty much a definition of what a hologram is. It is Vandervert's contention that the brain creates a hologram of an experience, and that perceiving any part of it recalls the entire experience, like Proust's *madeleine*. This is what causes the phantom limb experience. The hologram of the body is intact, a product of the brain itself. Vandervert then goes so far as to state that this concept can be extended to include "phantom seeing and phantom hearing experiences."[87]

A person born with only one leg may experience the missing leg from time to time as a phantom limb. The brain has created and maintained a holographic image of the body, and it contains both legs. Can this idea be extended, as Vandervert says, to include other "phantom" experiences?

If we can apply Karl Pribram's holonomic theory here, then we would have to suggest that experiences of alien abduction—"phantom" experiences, after all—would have to reflect real experiences that occurred at some point in the mind's development.

If the brain is a classical computer, as the advocates for artificial intelligence insist, these phantom experiences are not possible. There can be no paranormal world, no life after death, no mental telepathy because—quite simply—they do not "compute." However, if the brain is a *quantum* computer these things become not only possible, but probable. When we apply chaos theory to the quantum world we can begin to explain creativity and imagination: the explosion of ideas and connections between heretofore unconnected ideas and events that stem from a simple quantum event, perhaps the firing of a neuron or a series of neurons, the Orch OR effect of Penrose and Hameroff.

But are creativity and imagination "real"? Is the experience of the phantom limb "real"? Or do these events represent a kind of "twilight zone" between the classical world and the quantum world? Evidence of a mechanism in our brains that is reaching out beyond the confines of the skull to touch . . . what?

For most of us, these experiences are interior, internal to ourselves. They happen inside our heads. We can't demonstrate them to others unless we concretize these experiences in some way, such as by creating a work of art or writing a piece of music or a mathematical proof. They start within our skulls and work their way out if we let them. The hologram-like nature of consciousness flits *behind* our eyes, not in front of them.

But for some of us, there are experiences like creativity, like imagination—in other words, that share some basic qualities with these events—that come from *outside* of ourselves, moving *in*.

And then you have UFOs. And alien abductions.

The same mechanism that operates within our brains has been affected by something outside of it. The mechanism that gives rise to our creative impulse has been "switched on" by an outside force, a force that is using it to communicate with us. Forcing us to see things we ordinarily would not see. To dream while still awake.

Sounds spooky, we know.

But if all of these consciousness theories have taught us anything, it is that our neurons—and their associated microtubules—are operating at a level far beyond what we can consciously understand or appreciate, at a quantum level where non-locality and entanglement occur. Where consciousness itself occurs at the moment of the collapse of the wave function. A collapse that—theoretically—could be engineered from an external source.

We already know this to be possible. We know that drugs can affect the neural activity of the brain. Anesthesia can do that. Certain EEG-type equipment can do that, mess with your brain waves, make you laugh or cry. These are all clumsy methods of interfering with the firing of our neurons and the manipulation of our microtubules.

We are accustomed to seeing ourselves as discrete, separate components of a larger society, as individual and unique organisms with self-identity. If you want to interfere with our brains you need to get our permission first, and then use all sorts of exotic mechanisms or substances to do that. You have to operate on the individual person, focus your attention on the specific cranial anatomy.

That would be the only way to approach the problem if consciousness was contained within the skull like wine in a crystal goblet.

But what if it wasn't? What if your own, very personal consciousness was accessible from some distance away? No, we don't mean radio waves or ELF waves, although those are possible. What we mean is using a quantum process to communicate directly with the quantum processes taking place right now in your brain. *Mano a mano*, or microtubule to microtubule.

If the flapping of a butterfly's wings in China can cause a hurricane in Texas—according to chaos theory—then can a slight, quantum-level nudge from an outside source cause a consciousness hurricane in our brain? Is the experience of alien abduction or even of the UFO itself the result of just such a manipulation, a control mechanism for altering human consciousness by approaching it beneath normal awareness, at the quantum level, below even the level of the neurons?

If consciousness is holonomic, as per Pribram and Vandervert, then a tiny portion of that hologram is all that is required to re-create the entire image or neuronal event. Can that "slice" of the hologram be inserted into consciousness in such a way that we are aware of it not when it happens but only after the chaos effect takes place and it balloons into a full-blown "hallucination"?

Stanislas Dehaene is one of those cognitive psychologists at the forefront of consciousness research who is at odds with the Orch OR theory in that he claims it is based on "no solid neurobiology or cognitive science."[88] That opinion is surely debatable considering the role played by microtubules (which are, after all, an essential element of the neuron, which is the core of neurobiology) in the Orch OR theory. Dehaene believes that consciousness is an emergent property of the brain,[89] and thus quantum mechanics would play no part in what certainly was an evolutionary process. Regardless, one of the themes in the work of Dehaene is that of subliminal images and the way they reveal hidden cognitive function.

We react to subliminal images even though we are not consciously aware of them. This seems counterintuitive, but decades of research in advertising and media studies demonstrates the truth of the claim. When the film version of *The Exorcist* was released in 1973 there was a persistent urban legend that said subliminal images were used to create more horror than the visual images themselves. This turned out not to be true, but it raised awareness of the possibility during a turbulent time in American history, when rumors of intelligence agency manipulation of the media was at an all-time high.

A subliminal image is one that lasts for a very brief period of time, too brief to be noticed, but which the eye records and sends back down the optical pathway to the brain. It is subliminal in that it appears below the threshold (the *liminis*) of conscious awareness. There is some debate as to where this threshold is found, but according to Dehaene the basic assumption is valid. A strip of 16mm movie film is actually a series of still images which, when run at the appropriate speed, trick the eye into thinking it is observing motion. These images, or frames, run past the eye at 24 frames per second in order to achieve the "flickerless" perception of continuous, smooth movement. If one were to insert a different frame in the strip of film—with a written message or an image—it would not be noticed but would instead fall beneath the threshold of awareness. That image would still register in the viewer's brain, however, and potentially influence some action whose motivation would be unknown to the viewer on a consciously aware level, but known on a subconscious level.

Dehaene is a great believer in the importance of the unconscious mind, and claims that most mental processing takes place there with the conscious mind (located in the prefrontal cortex) as a kind of overseer or watchdog over the unconscious processes, organizing and prioritizing the data coming in from other parts of the brain.

In addition to the phenomenon of subliminal messaging there is also the fact of transcranial magnetic stimulation, or TMS. This is the deliberate targeting of specific areas of the brain by placing a magnet on the skull (or surgically inserting it onto the brain itself) and passing an electrical current through it. Depending on where the magnet is placed, it can produce emotional responses, auditory and visual hallucinations, even orgasms. Memories can be retrieved or new experiences created, such as the feeling that one is floating outside of one's body and staring down at it (via stimulation of a part of the parietal lobe). Dehaene is quick to point out that "this fact by itself does not directly speak to the issue of the causal mechanisms of consciousness."[90] In other words, there is correlation but not causation. It is not clear how stimulating parts of the brain to elicit specific responses is a solution to the hard problem of consciousness. To use our admittedly hoary example, we can access a computer to retrieve a file but that does not mean the computer is conscious of its contents or has integrated the file with other disparate elements of its memory to form a comprehensive idea. The emotional responses experienced by those undergoing TMS are free-floating and un-anchored, without context.

> "Indeed, recent research suggests that the initial bit of induced activity is unconscious: only if activation spreads to distant regions of the parietal and prefrontal cortex does conscious experience occur."[91]

Or, as Stuart Hameroff might say, when the wave function collapses.

The "bit of induced activity" may not be restricted to TMS experiments alone. This weird, externally created activity conceivably could be stimulated by a different source, something that would excite a neural response but remain unconscious until it reaches a certain threshold and the experience of it creates a cascade along the neural networks. This was the idea behind Proust's *madeleine*, for instance: the taste of the tiny pastry evoking a flood of memories. But in that case, Proust's protagonist had also tasted the confection earlier in life, and it elicited a number of memories that were associated with his childhood. There was a specific context. While TMS evokes memories by sending a current into the appropriate section of the brain, with or without context, is it possible that we have memories for which there is no conscious awareness? Memories that remain dormant, perhaps for our entire lives? And could those unconscious memories exert any influence over our conscious actions?

Of course, that is the claim found in the texts of depth psychology and appears in the writings of Sigmund Freud and Carl G. Jung, among many others. A childhood trauma, for instance, will be played out constantly in adulthood until it is recognized and neutralized through therapy. The experiences are real, but they are ghosts. They haunt the living just as their horror-story counterparts haunt old castles and cemeteries. In this case, the ghost is evidence of something real. It is the trace element of an old experience. We function—often badly—as a result of something dimly remembered, or not remembered at all. If the psychological or emotional disorder exists and there is no discernible organic cause, the analyst digs deeper. The "ghost" is proof that something was once very wrong.

What does all of this have to do with subliminal messaging, transcranial magnetic stimulation, and chaos theory? Further, what does it have to do with the Phenomenon?

If the Phenomenon exists, then it has to leave traces in the psyche of human beings. These traces may not be discovered easily, depending on how they were left. Sometimes a dream can seem so real that our memory of it outlasts memories of real events. And sometimes we can identify an experience as a dream even though we know that it wasn't, that it shares nothing in common with our ordinary dreams. Yet the unreality of it, the lack of social or physical context, insists to us that it *must* have been a dream.

When John Mack interviewed people who identified as alien abductees, he began to realize that they were suffering from a form of post-traumatic stress disorder (PTSD) that had no known trigger. They were not war veterans. They had not suffered childhood trauma. They had not been victims of violent crime. They were in all other respects normal, well-adjusted human beings. Their only "aberration" was their memory of an experience involving aliens, abduction from their homes or vehicles, lost time, and often some kind of mysterious but traumatic surgical procedure.

If we apply what we have been studying about consciousness studies so far, we will see that there is a mechanism in our brains that can serve as a medium through which these experiences of alien contact and alien abduction begin to make sense. We have to take the entire experience as a coherent whole before we can approach the reality of it. It is not just the experience of being abducted; not just the image of the traditional Gray alien; not just the surgical procedures; and not just the telepathic communication. It is all of these, taken together and explored for coherence with the structure and behavior of the mind.

The phenomenon of telepathic communication is the gateway to the rest. We will discuss telepathy in greater detail later in this book, but for now we can say that the transmission of images from the Gray to the abductee is a clue as to what kind of being the Grays represent. This is a direct, mind-to-mind contact that assumes some degree of similarity between the brain of the Gray and the brain of the human. If it represents a quantum effect of some kind, then a great deal of what we have heard about the Grays starts to pull together. Telepathic contact would stimulate areas of the human brain and human consciousness that are rarely, if ever, experienced in the average person. If telepathy involves a kind of "quantum entanglement," as Hameroff suggests, then it becomes possible that the experience of being abducted—of floating through space, of passing through walls—is a conscious approximation of the entanglement. We have already seen that stimulation of one part of the parietal lobe with TMS will make a person seemingly experience levitation.

This is not to say that the alien abduction experience is a psychological illness or the result of the misfiring of neurons, or something similar. Far from it, in fact. What we propose is that the alien abduction experience is real *in some way*: it is the ghost of an actual experience that involves an external actor or actors, beings whose ability to manipulate consciousness is without parallel, but an ability that human beings are soon to acquire. That these beings are ordinarily invisible to us, that they seem like genies or sprites or ghosts or devils, should compel us to ask deeper questions about the experience and not write it off as the product of an overactive imagination, or hysteria, or mental disorder.

In many cases, the first abduction experience is not remembered immediately. After the second or third experience there is a realization that it has happened before. The memories begin to form as real memories, memories of a real event or series of events. As we noted earlier, an unusual experience may not "imprint" on the conscious mind immediately because there is no context for it. An alien abduction would certainly fit into that category. But repeated experiences will create a pattern of neural firing that will be associated with them, and it becomes easier to "remember" them with time.

In a sense, those initial experiences are subliminal. They take place under circumstances that render them invisible or imperceptible at first. Unconscious. The alien abduction experience does not involve an intruder breaking into your bedroom and dragging you out of your home at gunpoint. This is a different kind of abduction, and we only use that word because there is no other term for it that seems to fit the sensation of being powerless, and under the complete control of strangers: a situation in which the victim has no sense of agency at all, no ability to refuse or reject, no safeguards against it happening again. There is no plotting to escape or evade; no stockpiling of weapons; no cutting a deal with a corrections officer or a terrorist. This is a different experience entirely, and it takes place under circumstances that can only be considered paranormal in nature.

Who are the alien abductors? Why do they seem to remain invisible most of the time? How can something so invisible cause so much change in a human being?

As mentioned previously, it is possible to insert a subliminal message in a strip of film, as a single frame in that film, and then run the film at the usual 24 frames per second. You won't be aware of the message, but it will make its way through to your unconscious mind.

What if you could slow down that film strip, though? What would you see?

Take the example of a deck of 52 playing cards. You shuffle the cards and play Solitaire. No problem. But what if a Joker is added to the deck? As you shuffle the cards, flip through them quickly, you never notice that the Joker is there. But as you play the game, you slowly come to the realization that something is wrong. You can never win. There is an extra card.

The Visitor—the Gray, the Alien—is the Joker in the deck. If we flip through at our usual speed, we will never see it. All we know is that something is wrong, some flaw in the process. And if we simply replace one card with the Joker so that the total is still 52 cards, the result is mental illness, confusion, depression. That's because we have the right total, but one of the four suits is compromised. The balance is off. The suit is incomplete.

Most of us stubbornly insist on playing our game of Solitaire with a compromised deck, and we keep hoping that eventually we will win the game. The alien abductees, on the other hand, have spotted the Joker.

Or the Joker has spotted them.

SECTION THREE

▼

HUMAN-MACHINE
SYMBIOSIS

INTRODUCTION TO SECTION THREE

Let me remind you of our cardinal principle: through conscious means we reach the unconscious.

– Stanislavski[92]

Language is the first and last structure of madness.

– Foucault[93]

AVING LOOKED AT HUMAN ANATOMY AND NEUROBIOL-ogy, genetics, the various theories of consciousness, and how all of this pertains to our study of the alien, we now proceed to the point at which we approach the subject from a different direction.

After all, the title of this project is Sekret *Machines*.

The scientists who are working with us are deeply involved in the study of brain-computer (or human-machine) technology. The implications are enormous, of course, and they range from cyborgs to some novel approaches to the "hard problem." These include the study of the paranormal: extrasensory perception (ESP), psychokinesis (PK), and the whole range of parapsychology.

The study of the paranormal is often derided as "woo" by skeptics and critics who claim that anything a psychic claims to do is nothing more than stage magic and fraud, but many scientists today beg to differ. Indeed, as we get closer to a definition of consciousness that includes the neurobiological substrate—whether of the neuron or the microtubule—we approach a theoretical basis for paranormal abilities that rests on a scientific framework. Whereas previously any discussion of the paranormal placed it outside of classical physics, which resulted in it being characterized as unscientific and invalid, today it has become the focus of study by a number of research groups and institutes in the United States and around the world.

One of the most common phenomena associated with alien contact or abduction is that of mental telepathy. Communication between "aliens" and humans seems restricted to the transmission of mental images: a nonverbal form of communication that suggests not only that the "alien" has the ability to transmit a message but that the human being can act as a receiver of this message. This implies a structure in the human brain or nervous system that already is capable of telepathic communication; otherwise, any such transmission from the alien would be futile. Another possibility is that—as some scientists believe—there is a "pan-psychic" element to the universe in which consciousness permeates all of creation and is not as dependent on the anatomical structures of specific brains and nervous systems as we like to think.

After we consider some of these possibilities and look at several of the personalities and programs that were designed to address them, we will finish this section by reviewing what we have learned and applying that knowledge to a brief analysis of the typical "Gray" alien. Can we make some assumptions about them? Is the experience of the "alien" the same across cultures? If not, what are the common denominators (if there are any)?

The purpose behind all of this is to demonstrate the complexity of the problem of stating unequivocally that the Phenomenon is "this" or "that." The Phenomenon resists all forms of English-language description, grammar, and vocabulary. Indeed, it resists all forms of any language, but the inclusion of non-English, non-Western ideas, terminologies, and philosophies in our multidisciplinary approach to the problem will reap much greater (and faster) rewards than would be the case for any one group of scientists acting alone or in relative isolation. This is, after all, a phenomenon that partakes of consciousness, and it is the consciousness of all of us that will contribute to the ultimate solution.

We are not naïve enough to believe that a race-neutral or gender-neutral approach can be found or even should be enacted; on the contrary, we feel that a race-inclusive and gender-inclusive approach is necessary, recognizing that different cultural and social experiences contain such a wealth of information that it would be self-defeating to ignore them. To homogenize human experience into a bland monoculture of scientific endeavor would be to rob us of the very inputs we all need: those that seem idiosyncratic to some of us but are normal to the rest.

By examining the idea of race, for instance, and recognizing that it is socially constructed, we can open our eyes to the reality of the alien: a being, a presence, and a force that is likewise being socially constructed in our culture, right in front of us. These constructions blind us to the reality of the idea under construction. The alien is real; it is something that exists. But we can't see it; or when we do, we project every human anxiety, suspicion, and hostility onto it. Sound familiar?

The alien has become the ultimate Other. But is it, really? How much of the alien is us? How much of its nature and how many of its characteristics do we share? Our feelings about the alien represent a kind of schizophrenic fault line in our society: the alien is either a benevolent space brother who is here to rescue us from ourselves, or a demonic being bent on our destruction. What does it say about us that we can't agree on the basic facts of what is probably the most important (if relatively invisible) element in our civilization? This is not merely a flaw in the scientific approach; it is emblematic of the human condition in general. It illustrates in sharp relief the reason why any foreign species would have second thoughts about revealing themselves to us in all their glory: they would receive from one element of human society the sweaty embrace of the tearful if somewhat professional empaths who want to be cradled and forgiven, and from the others a barrage of super-weapons deployed to kill anything that moves. We don't understand that we will do more damage to ourselves by these reactions than to any alien, but it is entirely possible that the alien understands this and has taken measures accordingly.

Yet, even these ruminations are anthropocentric. As humans, we have a hard time considering the nonhuman; a hard time thinking outside the "black box." Awareness and attention, remember. We are aware, now, of the alien presence, of the reality of the Phenomenon. But we have a hard time paying attention to it. We associate it with other things, other ideas, other images, and thereby have convinced ourselves that we know what it is.

If you want to get a glimpse of the alien, sit quietly in a darkened room. Think of all you have seen and heard about the alien, and gradually reject each one of these ideas. Deny these images their reality. Go through until there is nothing left but emptiness. Wait. You will feel something gradually assuming form around you or before you or behind you. Remain calm.

You've made contact.

Now, can you describe it for us?

Now, do you understand?

THE BRAIN-COMPUTER MODEL

... providing the means by which man may someday be able to program his personality, or its better aspects, into the deathless machine itself, and thus escape, or nearly escape, the mortality of the body.

– Loren Eiseley, *The Invisible Pyramid*

WE HAVE TOUCHED BRIEFLY ON ARTIFICIAL intelligence (AI) in this book, and now we have to examine it more closely to see what contributions AI can make toward an understanding of the Phenomenon.

▼　▼　▼

As we mentioned, there are two main schools of thought concerning consciousness, and one is that the brain is a computer and consciousness is an emerging property of the brain, a kind of secretion of the neurons, with the obvious conclusion that an advanced computer would begin to develop consciousness. That is what is known as "strong

AI." It predicts the coming of the Singularity: that moment when machines become conscious beings, partners with humans in a bizarre new form of evolution. Ray Kurzweil, the leading proponent of this view, has even published a book titled *The Age of Spiritual Machines*. The problem with this view is the conflation of terms like "intelligence" with "consciousness."

The other school of thought rejects the idea that machines could ever become conscious, as there is a fundamental difference between consciousness and computation; in other words, machines can be intelligent, but that is not the same thing as conscious.

A third theory, also popular in some circles, is called panpsychism, which claims that consciousness permeates the universe, and that all things are conscious to some degree.

Artificial *intelligence* may not be equivalent to artificial *consciousness* (can consciousness of any kind be considered "artificial"?), but the pursuit of "strong AI" will eventually prove the theory one way or another. This has implications for our study of the alien contact experience, for if we can determine the difference between intelligence and consciousness—between computation or intelligence and consciousness—we may be able to identify the true nature of the alien contact experience, *at least as it pertains to humans*.

The problems we have had so far in this endeavor have centered upon language. We have never really defined consciousness; in fact, we are a little hazy on what constitutes "machine" as well. The insistence of some philosophers that human beings are little more than machines is part of the problem. If this is true, then it is only a matter of time—and not very much time at that—before we blur the distinction between human and machine to the point where shutting off a machine may be tantamount to murder. As ridiculous as that sounds—the staple of science fiction and fantasy literature—it is nevertheless a disaster waiting to happen, with all the social, ethical, and legal issues it implies.

Are machines an extension of human consciousness? The telescope extends the reach of our eyes into space; the microscope into the invisible world around us. The telephone and the Internet enable us to communicate across vast distances and reach uncounted numbers of people. At this time, a human being can sit safely in her own home and see stars and planets in real time, speak to someone on the other side of the world, watch hundreds of television channels or stream a film, listen to music composed hundreds of years ago (or just last night), read a classic work of literature, take a photo and send it to millions of "followers," order a pizza to be delivered or a book to be shipped. The machine has extended the reach of that human being into vast distances from a chair in her living room. (And still we're bored!) But is the extension of our sensory apparatus—eyes, ears, even touch—the equivalent of an extension of our consciousness?

That person in her living room may wear a hearing aid. Glasses. An artificial limb. A pacemaker. An insulin pump.

At what point does the line between the human and the machine dissolve? Or between "alien" and machine, or "alien" and human? And does it depend on sensory organs alone?

▼   ▼   ▼

Does all the talk about alien-human hybrids mask a simpler reality: that the "aliens" are themselves hybrids of organic life forms and machines? The fundamental questions would therefore seem to be, what does it mean to be human? What does it mean to be "alien"? And perhaps the key to answering both questions: What is consciousness?

One may be forgiven for thinking that these are unknowable issues, fodder for speculation and not much more. But if we apply critical thinking to the problem we will see that some answers are accessible. We just need to ask the right questions.

Consciousness studies are at the root of all of this, because we are on shaky epistemological ground. We are being told to distrust our own senses, or the senses of witnesses and experiencers, even as we fine-tune our sense-gathering equipment to detect more and more incoming data in a project that is consistent with certain preconceived notions of what constitutes data and what constitutes senses.

We are told that Ufology is a field of belief and not knowledge; in other words, that it is more like religion than science. (This is an artificial, socially constructed difference, as we suggested in the first book in this series.)

By studying Ufology—and all its related phenomena— we have crossed over into an area of human experience that has been rejected by the dominant worldview. In fact, we are being told that experience is *not* knowledge. What has been the reaction of eyewitnesses, either to UFO sightings or alien contact? It's likely some version of "If we tell you what we know to be true, we will be shot down immediately by critics on all sides of the issue and our statements ridiculed and then vilified." One of the excuses often given for denying the experience of those who have had alien contact is that there is simultaneously a strong consistency among accounts of alien contact and a disturbing inconsistency in important details. The consistency means (to the critics) that experiencers have compared notes, or that their experience derives from a shared cultural medium such as television or movies. The inconsistency means that they are talking about entirely different things that have nothing to do with UFOs (or that they are not very good at comparing notes!). Regardless, the experiences themselves are deemed invalid; at least, the experiencers are not describing what they think they are describing, because the experiencers are not trained (scientists, engineers, military officers, intelligence officers . . . insert the appropriate category).

There is a missing vocabulary here, a set of symbols that can communicate to us what the experiencers and the abductees have seen or felt in a way that is comprehensible to those of us who haven't had those experiences. We— as humans in general, regardless of race, ethnicity, or the other categories we use to order our world—share a certain

basic worldview in which a chair is always a chair, no matter what word we use in our own language to describe it.
A child is a child. Death is death. We surround all of these
experiences with our own cultural attributes and values, but
in the end if I say "child"—and it is translated into another
language—it means something that is readily understood
by all of us, because we all have been children. We all have
seen children. We know what a child is. For some of us,
a child may be inconsequential until it is older; for others, a child is already a human being deserving of the same
respect we give to adults. But we all know what a child *is*.

This is not the same for the type of experiences that
qualify as UFO/UAP or alien abduction, close encounters, etc. There is no common vocabulary that is built in
to the human experience in a way that is cross-cultural
and easily understood by everyone as meaning the same
thing. Media sources have given us all sorts of options for
describing the experience, and they have run the gamut
from tall, blond-haired, blue-eyed Nordic beings who radi-
ate peace and light, to short, sinister Grays who manipulate
our organs for their own unknowable purposes—and those
descriptions are just from the United States. These are all
representative of projections of cultural expectations onto
something that is, at heart, ineffable and beyond quotid-
ian human experience. These images represent the intense
desire of a human nervous system in shock at confronting
something for which it has had no prior experience.

For that reason (and several others) we have been forced
to look deeper into what constitutes knowledge, memory,

and intelligence. To do that, we have to look at language as well. If a machine can exhibit these qualities, then how is a human different? And perhaps fundamental to all: why is a human being of any value at all in the long term? In turn, that forces us to ask, what part of a human being is irreducibly human?

▼　　▼　　▼

As we discover physical laws, we should realize that these laws must apply to the Phenomenon as well, at least in some way if not in an identical manner as they do to us. If we can define consciousness and understand how it works, this understanding should apply to the Phenomenon as much as it does to us. If the consciousness of an "alien" is not the same as ours, it still must exhibit some qualities in common with ours.

According to cognitive scientist and philosopher David Chalmers,[94] consciousness should be considered "a fundamental feature, irreducible to anything more basic." In other words, we should view consciousness the way we do space, time, and mass: as a basic element of the universe.

It was Chalmers who coined the term "hard problem of consciousness" as a way of identifying the feeling that is associated with sensory data and asking why that feeling, that subjective experience, even *exists*. What purpose does consciousness serve? If we were truly machines, all we would need to do is process information, produce whatever it was we were designed to produce (by whom, why, and for what?), and take appropriate actions where required. There would be no need, for instance, for entertainment. Music, literature, the arts . . . all useless to a machine. We may say, with the materialists, that consciousness is an emergent property of the brain, but then how is *Moby Dick* an emergent property? Or a Valentine's Day card? Or a funeral service? Ray Kurzweil tells us of computer programs that write poetry and prose; other researchers— such as Ian Goodfellow, who works with the Google Brain Team—have been working to give machines "imagination" so that they can produce works of art.

Goodfellow's approach[95] has been to create something called a "generative adversarial network," or GAN. The basic idea is to have two networks fight with each other over the solution of a problem. Problems that would ordinarily take days or weeks to compute can be solved by dueling neural networks in a fraction of the time. In the GAN network there are two main components: the generator and the discriminator.

The generator creates a stream of artificial outputs. These can be images of persons, animals, objects, or virtually anything else.

The discriminator then matches the output of the generator against a database of original images and in so doing determines which of the generator's images are false and which are genuine. These go back to the generator.

The generator then creates another output stream based on the input from the discriminator, and tries to fool the discriminator into selecting one of its false outputs as "real."

Eventually, the discriminator can't tell the difference between a real image and an artificially generated image.

The generator wins.

This technology is now being applied to a range of video and computer games, for obvious reasons, as well as Google Maps and other apps, but its real contribution may be to medicine, particle physics, and other disciplines that usually require months or years of running simulations and tests that could be accomplished much more quickly with this approach. In fact, engineers are now looking at GAN as a way to provide machines with consciousness. Such a system would be able to "self-learn" without being taught by a human AI programmer. Once the system is in place, the machine would conceivably run through simulations of virtually anything you set before it and—by comparing simulations to other datasets, including those culled from a massive mining of the Internet—it could learn foreign languages, write poetry, compose music, and perform many other functions as a product of what AI specialists call "deep learning."

The downside, though, should be obvious. Any system that can create perfect simulacra of information—fake images, fake text, etc.—can also create "fake news," especially because "news" is a combination of . . . text and images. What we complacently refer to as "virtual reality" could so easily become "fake reality," and decisions based on that reality could have tremendous consequences for the rest of us. There are already concerns being expressed that self-driving cars, for instance, could be fed perfectly credible images of streets where none exist, or read a street sign incorrectly if it has been defaced in some way. The learning curve could be steep, but it is a Catch-22 situation: whether the system is flawed or perfect, it has the potential to cause social, political, and cultural dislocation on a scale never seen before.

One example that came to light recently was the sudden appearance of fake pornography featuring famous people in the starring roles. It developed that the technology is relatively simple, once the software—which is readily available—was applied to the problem. (A similar technology was used in the *Star Wars* film *Rogue One* in order to have Peter Cushing "appear" even though he was deceased.) The female lead in the *Harry Potter* films—Emma Watson—was one of those depicted that way. It's called "deepfake celebrity porn" and it surfaced on Internet sites such as Reddit in the last months of 2017. In basic terms, it consists of superimposing the face of a real person over a generic, 3D-created face. The newly created face is not merely a static object but is complete with facial gestures and eye movements to the extent that it becomes difficult to tell that you are looking at a simulation and not the real thing. In this case, the real face (Emma Watson's) was superimposed over that of a porn star so that it appeared as if Ms. Watson was actually performing in the film. Sites carrying the films were eventually banned (even by porn sites) and the films themselves submerged beneath the waves of the Dark Web.

But the damage is done. Pandora's Box has been opened, and the full implications of what this new technology can do are exposed for all to see.

It's often said that the first casualty of war is the truth. Similarly, the first casualty of AI is trust. It is gradually dawning on us that the manipulation of digital reality is pervasive and increasingly undetectable. We simply can't trust what we see any longer. It's a cynical reversal of the old Groucho Marx line, "Who are you going to believe? Me, or your lying eyes?"

Aside from the obvious and deliberate manipulation of digital media to promote a political position, support (or create) a conspiracy theory, sell products, or defame an innocent person, there are reasons to be worried about the explosive growth of AI that rest in the unspoken and unacknowledged biases that are inherent in the process. These include endemic racial and gender attitudes of which the AI engineers may not even be aware.

Bryor Snefjella, a PhD candidate in the cognitive science of language, on March 22, 2018, published an article in which he warned about the weaponization of AI and the potential use of the latest natural language understanding and generation research to promote authoritarian regimes. He takes that concern one step further:

> One widely used technology that enables machines to measure the meanings of words has been used to reveal how artificial intelligence systems can inherit the racial and gender biases of their training data.[96]

(Even this book is subtitled "Man." Does that mean there is an inherent gender bias in what we are trying to say, or is it a more blatant expression of a deeper mystery: the recognition of a deliberately gendered approach to the Phenomenon? How would this book be written or organized differently if the subtitle was "Woman" or even "Human"? Or if it had been written by a woman? Or a person of color?)

Just a thought: what if the AI systems running in our putative "aliens" have the same problem hardwired into them? What if certain biases—perhaps not race or gender as we understand them, but something even more inimical, such as biases against organic creatures, or creatures that sexually reproduce, or are taller than they are, or have a sense of humor—are integral to their programming; essential components, as it were?

Not a bug, but a feature.

▼　　▼　　▼

When we read or hear of the experiences of alien abductees, we get no sense that the alien environment has art, literature, or music. The alien environment appears utilitarian and lacking in cultural artifacts. This has led some people to suggest that the alien—particularly the archetypal "Gray" alien—is not an actual entity but instead represents a mental projection, a *human* cultural concept of the ultimate intelligent being (which is not saying much about how intelligent beings are viewed by humans!).

We remember the character of Mr. Spock from the *Star Trek* television series as logical, emotionless, and humorless. He was anatomically similar (virtually identical) to human beings, however, and was not a short Gray. But his affect was closer to a Gray (as they have been reported) if Grays could speak as we do. Spock, however, was not artificially intelligent; his intelligence—superior to human intelligence—was a product of his genetics and his planetary environment. The Grays are less anatomically similar to humans, but as mentioned previously, they do possess some of the basic anatomical features that are markers for intelligence and consciousness. The question then is very basic: are the Grays—are the aliens in general as experienced by abductees and others—artificially intelligent, or even artificially conscious?

Perhaps our consciousness is, as Giulio Tononi[97] would have it, the result of a process of integrating information. In other words, our brain integrates the data coming in through our senses into a composite picture of reality. For that to happen we need fully functioning sense organs and pathways from the sense organs to the neurons in our brains. If the sense organs are impaired—or nonexistent—there is less information to integrate, and a corresponding impairment of full consciousness.

One implication of this theory is that a plant, for instance, may be conscious. Its system integrates data coming in—via photosynthesis, but including water, temperature, location, nutrients in the soil, pollutants or other chemicals in the air—just as a human brain does.

The difference of course is in the lack of other sense organs (hearing, for instance). If the plant can hear—as some theorists insist, claiming that playing music for a plant increases its health and growth rate—it still cannot speak. The lack of an organ for speech must affect the consciousness somehow even though speech is not a sense in our usual understanding of the senses.

The Grays—to use our common example—do not speak, either. Like the golem of the Prague ghetto, they are mute. They also have no ears. It would seem that sound and speech in their case would be a deficit in their consciousness, whether they were organic beings or machines. Their consciousness would not be able to integrate that type of data. If they cannot hear, then human sounds such as music, different languages, etc., would be lost on them and would not be included in their understanding of *our* consciousness. That would reduce their experience of humans by at least one full sense.

The lack of a functioning oral cavity would seem to suggest that the Grays also do not eat. This might be an assumption, but they have gone to some lengths to look like humans and to have a kind of vestigial mouth, but they do not seem to use it for anything. The experience of eating—and eating the way humans do, with our diversity of cuisines and the flavors, tastes, and aromas associated with them, an essential aspect of culture—likewise is lost on them. If they do not understand diversity of sound and diversity of cuisine, then how much of us can they really comprehend?

This leads us into another discussion, and it concerns artificial intelligence.

As we build more and more sophisticated machines with vast computing capabilities, gradually leading us to the Singularity, to Convergence . . . will we equip those machines with appetites? Will they eat? Drink? Will they be able to appreciate a good cup of coffee or a rare cognac, be able to tell the difference between Chinese and Japanese cuisine, between Spanish and Italian? Not just on the basis of terminology or ingredients, but by taste? Will a machine prefer one taste to another, be able to tell if a dish was made the right way?

Will a machine be able to taste a *madeleine* and have a Proustian moment?

If they don't, can they still be said to have a consciousness the same as ours? If it is different in critical areas, will that pose a problem for us? Will that make them dangerous?

Doesn't Genesis tell us that at one point God wanted to destroy his creations, the "machines" he made in his "image and likeness"?

Regardless of the answer to that question, there is no doubt at all that we are approaching the Singularity. The pace of technological development can lead in no other direction. When full human-machine integration—human-machine symbiosis—is achieved, will we be able to distinguish ourselves from the Grays?

If our symbiotic devices have no need to eat or drink, we will have reached not only the Singularity but a different kind of Convergence, one that would take place between humans and aliens, and one that would presumably obviate the need for sexual reproduction, thus eliminating another aspect of "human-ness" and the source of much of our art, spirituality, and culture. Sexuality would become a curiosity, a memory of a more savage time perhaps, evidence of the emotional vulnerability of human beings, a bug in the software, something to be fixed in Human Being Rev. 2.0 for the benefit of the machines. Such a symbiosis would result in humans losing something, some aspects of their identity as human, and the machines gaining a great deal. That seems to be an element of arguments in favor of strong AI: that we would rid ourselves of things we don't need, of extraneous issues that do not contribute to the greater good (as defined by whoever it is that is put in charge of technology). We would become more efficient, less wasteful . . . in fact, more like the machines we are creating. This would enable us to travel to the stars more economically (in terms of time, money, and energy resources) and to colonize other planets.

But who would be doing the colonizing? Who—or what—will we have become by that time?

One of the criticisms of the UFO Phenomenon that is raised frequently is the observation that the aeronautical characteristics of the craft seem to argue against their being piloted by organic beings. To fly at tremendous speeds, only to stop on a dime or perform a 90-degree turn at velocity, would create enormous G forces that would destroy a normal human being.[98] Thus the speculation is raised that perhaps the "pilots" of these craft are not organic, flesh-and-blood creatures at all, but robots. Cyborgs. Machines. Such a possibility would also explain the machine-like (and seemingly sexless) nature of the Grays.

Is this a goal worth reaching, this Convergence of human with machine? It would enable us—or something like us—to reach the stars; to "boldly go" where no human has gone before. And as we boldly go, we will be leaving something—maybe everything—behind. Not just the planet Earth, but our Earthly selves as well.

The closer we get to strong AI, the closer we get to identifying with the aliens; both those of our dreams and those of our nightmares.

▼　　▼　　▼

As we look at some of the technologies we have been developing to probe the mysteries of the human mind we should always keep running, on a parallel track, an awareness of the implications of our research and development where alien existence is concerned. As we make greater and greater progress, we will uncover strengths and vulnerabilities in our technology and in our neurobiology that will raise important questions about our ability to control various outcomes. The closer we get to blending what it means to be human with what it means to be a machine, the closer we *may be* getting to understanding what it is that has been visiting us since recorded history. The closer we get to a perfect AI system, the closer we approach the ultimate existential crisis where human (and alien) consciousness is concerned.

This is not an attack on AI, but a suggestion that maybe we are not thinking this through. AI will generate tremendous benefits for humanity, but there will always be a dark side to the Force. We know that Internet sites such as Facebook and Google have been accused of participating in a violation of our privacy as individuals, compiling our personal data and renting it out to commercial firms and political lobbyists, and even of challenging what we believe about privacy and making us wonder if privacy even still exists or is nothing more than a quaint notion held over from the twentieth century. This is ironic, because our concept of the alien—whether as a starship that appears anywhere out of nowhere, to hover over our landscape as a kind of "all-seeing eye," or as an abduction from our beds in the middle of the night—is tightly bound to the idea of privacy and the ensuing violation of our privacy that the seemingly omniscient alien presence implies. Even further, as we wonder about military or government knowledge of the Phenomenon and how many of us believe such knowledge is being withheld from us, we are forced to ask what the difference is between privacy and secrecy; the privacy of the individual as against the secrecy of the organization. We are told by some in the UFO community that we live in a surveillance state. Is such Orwellian surveillance an extension—an intrusion into the body politic—of a kind of "Alien Nation," an adoption by the state of the mechanism of alien observation of humanity, or has the fact of government surveillance merely suggested the "alien" narrative as a way of camouflaging government agency? When dealing with "strong AI," will we ever be able to tell the difference?

Trying to stop the march of technology toward the Singularity would be tantamount to burning the Library of Alexandria. It would be to deny knowledge and the acquisition of knowledge. We thus need to know ourselves—who we are, what we want, and if we want to survive as *humans*—before it's too late, before the machines make up our minds for us.

I, CYBORG

In the man-machine systems of the past, the human operator supplied the initiative, the direction, the integration, and the criterion. The mechanical parts of the systems were mere extensions, first of the human arm, then of the human eye. . . . There was only one kind of organism–man–and the rest was there only to help him . . . however, we see that, in some areas of technology, a fantastic change has taken place during the last few years. "Mechanical extension" has given way to replacement of men, to automation, and the men who remain are there more to help than to be helped.

– J. C. R. Licklider, "Man-Computer Symbiosis," *IRE Transactions on Human Factors in Electronics*, March 1960

THE QUOTATION THAT BEGINS THIS CHAPTER IS TAKEN from a groundbreaking work on man-machine symbiosis, published in 1960 (nearly sixty years before this book is being written). It was Licklider—working for the Pentagon—who developed one of the first versions of what would become the Internet, calling it the "Intergalactic Computer Network." From Licklider's computer networking concept of the early 1960s grew the ARPANET of the late 1960s and eventually the Internet as we know it today. His memorandum on the subject, dated April 23, 1963, was addressed to "Members and affiliates of the Intergalactic Computer Network" under the aegis of the Defense Department's Advanced Research Projects Agency (ARPA).

Already, in 1960, Licklider noted that it seemed as if men were there to help the machines rather than the other way around. He was referencing automation, which at that time was all the rage in engineering and manufacturing circles. We have come a long way since the days when a computer took up an entire building and had to be kept cold, and the human interface was by way of punched paper cards or punched paper tape. (One of the authors—you can guess which one!—remembers using just such a system in the 1970s when he worked for a multinational manufacturing firm. Thus, in his own lifetime, he went from using punched tape and keypunch cards to communicate with a computer the size of an entire room to the keyboard and mouse arrangement he is using now to talk to a computer on his desk that is many times more powerful than any computer that existed in 1975.)

Licklider worked for the Pentagon, as noted, and therefore his focus was on developing systems that would enable the military to make quick decisions concerning matters of defense and national security. His background was in psychology, physics, and mathematics. He migrated to MIT (1950) from Washington University in St. Louis and later Harvard, and eventually to ARPA by 1962.

We will discuss ARPA and related organizations in more detail in *Sekret Machines: War*, but for now it is useful to mention that ARPA was created in 1958 by order of

President Dwight D. Eisenhower in response to the Soviet launch of the world's first satellite, Sputnik 1. He realized that the Russians had utilized a broad array of scientific disciplines in order to accomplish this feat, and that the United States had to do the same. This was not the first time such a program was proposed, however. Earlier, at the end of World War II in Europe, General Hap Arnold asked the famous scientist Theodor von Karman to investigate how the Germans had done something similar with their rocket program at Peenemunde and later Nordhausen. Subsequently, von Karman—after spending time in Europe as the war was winding down and interrogating Nazi scientists and engineers—came away with a proposal for an integrated, multidisciplinary program similar to what later would become ARPA.

In 1958 Eisenhower also created the National Aeronautics and Space Administration (NASA), which took over many of the duties of ARPA. This relegated ARPA to a newly defined mission involving the investigation of basic research into unconventional areas, those otherwise considered too high risk or perhaps not likely to result in immediate progress. It was this freewheeling atmosphere of "thinking outside the box" that presumably lured Licklider (and others like him) to ARPA.

National security was still the main focus of ARPA, and it included new approaches to the problem of missile defense. While still at MIT, Licklider was a key player in the development of the Semi-Automatic Ground Environment (SAGE) system. This was a system in which computers

299

collected data but presented them to a human operator who then had to make a decision as to how to react to the data. Licklider's contribution was on the human operator side, and this led him to investigate the relationship between humans and machines, specifically computers but including other machines involved in automation, and from there to develop his theories concerning "man-computer symbiosis." His approach was not so much AI but IA: i.e., intelligence amplification. He suggested that machines would (or perhaps should) extend human intelligence rather than replace it. In the end, it would be a human who would make the important decisions in any system, using the machines—computers, process control instruments, etc.—as tools or as sensors. This concept seems to have been abandoned as computer scientists realized that there was a lot AI could accomplish. Why stop at the computer as a mere sensor or computing tool when it could do so much more?

What did remain of Licklider's concept, however, was the idea that a "symbiosis" was possible, even desirable, between humans and computers. It is possible that he felt consciousness was not going to be achieved by machines, and that humans would still hold the upper hand in the relationship. By the beginning of the twenty-first century, however, it began to appear that deeper human-machine integration would result in a blurring of the lines between them.

Before that would happen, however, other scientists were imagining a world in which humanoid creations would actually replace humans in some industries, and for specific purposes.

Enter the cyborg.

▼ ▼ ▼

The same year that Licklider published his article on man-machine symbiosis, authors Manfred E. Clynes and Nathan S. Kline published an article titled "Cyborgs and Space," in *Astronautics*. It contains the following, rather startling, opening sentence:

> Space travel challenges mankind not only technologically but also spiritually, in that it invites man to take an active part in his own biological evolution.[99]

Wow.

This linking of science to spirituality would be short-lived in the culture, but in 1960 it was not unusual to hear both terms in the same sentence, or uttered in the same breath. In July 1952, the US Air Force had made some of the same associations in a press conference held in response to that year's UFO overflights in Washington, D.C., linking the UFOs to Biblical events, spiritualism, and mental telepathy.[100] For a while, it seemed as if there was a "kinder, gentler" side to the most advanced scientific effort known to planet Earth. There was an acknowledgment that space travel would have an effect on human beings that was spiritual, but also that it meant humans would be taking "an active part" in their evolution; something with which many mystics and gurus would agree.

Clynes and Kline go on to state, taking respiration as an example, that different orders of being have different ways of "breathing" or taking in oxygen:

> Mammals, fish, insects, and plants each have a different solution with inherent limitations but eminently suitable for *their field* of *operation*. Should an organism desire to live outside this field, an apparently "insurmountable" problem exists.[101]

They reflect that humanity is in the unique position of being able to consciously solve this problem by developing strategies that enable humans to breathe in environments where human respiration would be impossible. For instance, we developed the apparatus necessary to enable us to breathe underwater. The same idea was used to enable pilots to breathe at high altitudes. Thus, humans were able to penetrate the oceans and the skies in relative safety, whereas a bird (for instance) can't breathe underwater, and a fish can't breathe in the upper atmosphere. This was the beginning of a man-machine symbiosis on a different level; not only in the realm of computation (and, by extension, consciousness), but extended to biology.

Clynes and Kline identify the basic elements of a strategy for this biological symbiosis, which is the creation of a self-regulating system that "must function without the benefit of consciousness in order to cooperate with the body's own autonomous homeostatic controls."[102] This is an interesting statement and one that deserves closer scrutiny.

Consciousness is characterized here as a kind of awareness and attention (as we saw earlier). Yet we know that many of the body's systems operate below any kind of conscious awareness of the "owner" of the body. We breathe, our hearts beat, etc., without any type of conscious intervention. In order to facilitate space travel, the authors contend, a kind of self-regulating system must be developed to take over operation of these unconscious controls since the astronauts would no longer be on Earth, where these unconscious systems operate freely. We breathe on Earth without a second thought, but in space that is not possible. Therefore, what our bodies do unconsciously will have to be imitated by machines that enable us to breathe. In other words, a machine-derived "unconscious" must be developed.

By posing these questions, and by designing these systems, we may be closer than ever before to not only answering the consciousness question but also of understanding how a being that is not human could behave like a human, seem human-like, and function with autonomy like a human being, and yet still be completely alien to us in ways we intuit but cannot explain.

Clynes and Kline have an app for that:

> For the exogenously extended organizational complex functioning as an integrated homeostatic system unconsciously, we propose the term "Cyborg."[103]

And we are off to the races.

We learn in a more expanded version of "Cyborgs and Space" that it was Manfred Clynes who came up with "Cyborg." It was a term that caught on, with all its implications of something that was not completely human and not completely machine; not a robot, because a robot is a machine, a simulacrum of an organic system. Cyborg is . . . well, something else entirely.

A cyborg—"cybernetic organism"—is a human being with implants or other modifications that enhance certain capabilities or provide alternate ones. Strictly speaking, one could say that a person with an artificial limb is a cyborg. This does not mean an android or other completely artificial human being, but a human being with artificial components. The authors even extended that definition to include the use of psychoactive or other drugs designed to enhance performance, survive long periods in space travel, and protect against psychological stress.

Cybernetics is a term coined by Norbert Wiener in 1948 as a communications concept, "or control and communication in the animal and the machine," which was the subtitle of his famous and influential text *Cybernetics*.[104] This was in the immediate postwar period, the same era that saw the beginning of the modern UFO experience as well as the start—in 1950—of serious attempts by government agencies and their subcontractors to probe the mysteries of the human mind in an effort to counter the effects of "brainwashing." There was the idea that consciousness could be manipulated directly, a realization that grew out of the psychological warfare and propaganda efforts on all sides during the war. One manifestation of this was the creation of the "Mad Men" advertising agency milieu, which was based on the wartime experiences of men like C. D. Jackson, who had worked for Eisenhower during the war as a propaganda specialist and then returned Stateside to become head of Time-Life.

Another manifestation was the scientific community's excitement over the possibility that the brain could be redefined not as the soul of a human being but as its central processor: a machine that could be tinkered with, manipulated, even perfected. Behind the mathematical formulas and quiet confidence of *Cybernetics* lay what some believed was a more sinister perspective: the quantification of human beings as little more than fancy machines, "wet" computers, or cogs in a vast "clockwork orange."

Clynes and Kline understood that Wiener was on the right track, but hastened to insist on a more spiritually inclusive approach that did not deny a person their humanity just because their brain could be likened to a fancy computer. The race to space was a top government priority in the aftermath of Sputnik, and there had to be a way to enhance human capabilities in space without sacrificing any component of what made a human being, human. That was in 1960. By 1963, the term "Cyborg" had become ubiquitous in the literature of a scientific establishment that was seeking to outperform the Russians in the space race. An article published by the United Aircraft Corporate

Systems Center in Farmingdale, Connecticut, that year states unequivocally:

> The Cyborg study is the study of man.[105]

This was a study commissioned by NASA's Biotechnology and Human Research division, and was submitted by "Robert W. Driscoll, Cyborg Program." It seems to be based on the original concept by Clynes and Kline, but without any mention of those two authors or their work. It covers much of the same ground, and indeed Clynes and Kline made a more formal proposal of their concept in an article published in 1961 as part of a collection titled *Psychophysiological Aspects of Space Flight*. In this paper, they were even more philosophical:

> Man must first conceive that which he would create. . . . We can only communicate with each other in the most inarticulate fashions, and we do not know how to create life. Our presumption is even greater than our ignorance: when we cannot understand something we call it UNKNOWABLE MYSTERY or ULTIMATE. The Epimetheans among us then cry "sacrilege" when the problem is even approached. As if God's infinity were but a finger's grasp beyond our own limitations![106]

This is from the opening paragraph of a paper published in a *scientific* study of space flight. (For those not up to date on their Greek mythology, Epimetheus was the brother of Prometheus. His name means "afterthought," as opposed to the "forethought" of Prometheus.) The paper itself is titled "Drugs, Space, and Cybernetics: Evolution to Cyborgs." (Sounds like a party! Well, maybe on Venus.)

Drugs were being considered as essential tools for extended space flights for a variety of reasons; first and foremost, for medical conditions that could crop up during long periods in zero gravity. For this, they pointed out the applicability of a certain type of osmotic pump that could be inserted into the body to supply "biochemically active substances at a biological rate." As an example, blood pressure could be monitored on a continuous basis and when pressure fell above or below a standard norm a chemical compound could be released by the pump to automatically correct it.

Then, they get into more intriguing areas.

> How are we to set the upper limits of "natural" human physiological and psychological performance? We can take as minimal the capabilities demonstrable under control conditions such as yoga or hypnosis.[107]

We should keep in mind that this paper was published during the heyday of the CIA's programs designed to explore the possibilities of mind control, behavior control, and even the paranormal. Hypnosis was just one of those areas under intense scrutiny by the agency's subcontractors.

One of the cyborg programs seems like it would have been a perfect fit:

> We are presently working with a new preparation which may greatly enhance hypnotizability, so that pharmacological and hypnotic approaches may be symbiotically combined.[108]

In other words, they were working on a drug that could be "pumped" into the system to enhance the experience of being under a hypnotic trance. Why would this be desirable?

> At such institutions as the yoga colleges in India, the human imagination is stretched by the muscular control of which even the average undergraduate is capable. Hypnosis per se may prove to have a definite place in space travel, but there is a prior need for much more information about the phenomena of dissociation, generalization of instructions, and abdication of executive control.[109]

That is a paragraph that could have come—with only slight modification—from any of the reports generated by MKULTRA. The inclusion of "dissociation" is especially interesting in this context, as dissociative personality disorder (DID)—the modern equivalent of multiple personality disorder (MPD)—was one of the areas being explored by the CIA's psychiatric subcontractors.

The report is well worth reading in full, even if some of the science is dated by today's standards. An interesting further elucidation of the possible role of hypnosis is described, and it makes for compelling reading. It has to do with vestibular function, which is the system that is involved with balance and with spatial orientation, and thus involves vision and movement as well. The authors claim to have already been experimenting with hypnosis to see its effects on vestibular function a full fifteen years before this report was written, and thus in 1945 just as the war was winding down. For these experiments they used a form of Barany Chair, which is a swivel chair of the type used for pilot training. These experiments test for nystagmus, which is the rapid eye movement one gets when spinning around an axis and then suddenly stops:

> Under hypnosis, it was possible to induce the nystagmoid reactions characteristic of acceleration and deceleration with the subject in a stationary position. Despite a trance extending to a depth of negative hallucinations, it was impossible to eliminate nystagmus while the patient was accelerating or decelerating. It therefore appears unlikely that hypnosis would eliminate the subjective clues produced by the vestibular organ.[110]

In other words, they were able to *induce* the sensation of spinning in a test subject who actually was stationary, but were not able to *eliminate* the sensation in a test subject, under deep hypnosis, who was actually spinning. That would seem to indicate that it was possible to induce the feeling of a physiological reaction in its absence but not possible to eliminate the feeling of an actual physiological response that was taking place. This suggests that there are differences in conscious experience, and while an experience may be duplicated or simulated, there are actual physiological experiences that cannot be suppressed or eliminated, at least not under hypnosis.

But . . . what is a "depth of negative hallucinations"? Unfortunately, the authors offer no further clarification.

The report also describes sensory deprivation and the need for further research and study as it applies to space flight. They suggest that it is not the absence of sensory stimulation that is the problem, but its *invariance*, specifically what they term "action invariance" or "action deprivation." They give as an example a space flight experiment (details withheld by the authors for some reason) in which two test subjects developed "marked visual hallucinations" even though they were able to move about the simulated space vehicle, were in radio contact with the test controller, and were even in visual contact via closed circuit television. It was the constrained space they were in, plus the fact that nothing inside that space ever changed, that contributed to the hallucinations. It was the invariance of the permissible range of action that made them see things that weren't there.

They close their study with reference to "Erotic and Emotional Satisfaction," or perhaps the lack thereof, and the possibility that there exists in the human brain a "pleasure center" but with no recommendations as to how that would be addressed in space flight (perhaps putting that osmotic pump to use, generating psychoactive chemicals?), and then with an acknowledgment of the possibility that psychoses could arise as a result of long-term space voyages. Should that occur, it would be possible that someone suffering from a psychotic break would refuse to take the appropriate medication, in which case there "should be a release device capable of being activated from the earth station or by a companion if there is a crew," according to Clynes and Kline. A mysterious and rather ambiguous recommendation, but one whose implications seem clear enough.

They end by stating:

> It is proposed that man should use his creative intelligence to adapt himself to the space conditions he seeks rather than take as much of earth environment with him as possible. This is to be achieved through the Cyborg, an extension of organic homeostatic controls by means of cybernetic techniques. . . . It is suggested that such existence in space may provide a new, larger dimension for man's spirit as well.[111]

Once again, this insistence on incorporating spiritual ideals within the purely scientific mission of space travel. The psychological aspects of space travel may have been what prompted these sentiments. Or was it the influence of Manfred Clynes himself?

I'm with you in Rockland
where the faculties of the skull no longer admit the
worms of the senses
—*Howl*, Allen Ginsberg, 1955–56

Born in 1925 in Vienna, Manfred Clynes is a world-renowned concert pianist, as well as a scientist and inventor. He and his family fled Austria in 1938 to escape the Nazis, and wound up in Australia. At the age of fifteen, he developed an inertial guidance system that actually contributed to the war effort during World War II. He was a frequent guest of Albert Einstein and played the piano in his home. One of Clynes' many specialties—developed over decades of research and investigation—is the relationship between music and science, and the measurement of brain signals. His scientific accomplishments have been enormous, and they are much too numerous and varied to detail here. After winning countless awards for scientific accomplishments he returned to his first love—music—and studied with Pablo Casals, among others.

His co-author, Nathan S. Kline, was a pure scientist who happened to meet Clynes by chance in 1956 and offered him a job as chief research scientist at the Research Center of Rockland State Hospital. Rockland State was a sprawling (some would say "notorious") mental hospital in upstate New York, and Kline was interested in pharmacology as a treatment protocol for depression and schizophrenia, although electroconvulsive therapy and other extreme measures were also employed. Kline is acknowledged for his revolutionary development and use of tranquilizers and antidepressants in psychiatry. The year Manfred Clynes accepted the position with Nathan Kline is the same year Allen Ginsberg completed his epic poem, *Howl*, which contains repeated references to Rockland. One wonders how the close proximity to thousands of mental patients affected Clynes' own thought processes. If we reread Ginsberg's poem, especially the last section, which deals specifically with Rockland, we are forced to consider that maybe there was a link, however tenuous, between the inner world of the schizophrenic patient and the imagined inner world of the lone space traveler.

No matter the underlying cause, Clynes began to devote his attention to the possibility of nonverbal communication. He began to wonder if maybe a human brain could communicate with another human brain without the messy stuff in the middle: sounds, words, gestures, inelegant phrasings, and misunderstandings. No syllables. No phonemes.

In later years, Clynes focused his attention on the miracle of the human eye. Like many of the philosophers of consciousness we have already discussed, beginning with Descartes, Clynes believes that vision remains the most promising of all the senses for understanding how the brain, and by extension the mind, works. Clynes, however, has a unique perspective. It involves the lens of the eye, which, he says, really doesn't belong to the body (it has no blood supply going to it, and it sits in a liquid by itself) and is therefore a kind of cyborgian implant already. We don't consciously control the lens; there is no feedback loop between the lens and any other system. Yet, the lens curves as required by the mind to see objects far away or close by.

> If we could tap the system that controls the lens
> to control something else, it would be "the nearest
> thing to telekinesis," as Clynes put it.[112]

So, lurking back behind all the science and the math, the physics and the biochemistry, and even behind the space program itself, is the ancient dream of the mystics and the gurus: a spiritual regeneration of the human being, but one that would include implants and prosthetics, voiceless communications and telekinesis.

As stories began to proliferate about aliens landing in spacecraft and communicating with their minds, our scientists were working and dreaming of just . . . precisely . . . that. Chicken . . . or egg?

CAN YOU HEAR ME NOW?

It CAN be said that IF telepathy exists, then it would be of such overreaching and extraordinary importance that all Earthside institutions would have to be "reorganized" in the face of it.

– Ingo Swann[113]

No progress can be made in our knowledge of UFOs without changing man's brain. And I'm talking about a biological change, not just a spiritual or psychic change.

– Aimé Michel, as quoted
in Jacques Vallée[114]

R UNNING PARALLEL TO THE PURELY SCIENTIFIC EXPLO-rations of consciousness, genetic engineering, quantum mechanics, and artificial intelligence is a study conducted more quietly but with no less vigor: the field of parapsychology. Like the serious analysis of the UFO/UAP Phenomenon, this study is carried out in relative secrecy (or obscurity).

The logical, reductionist, and materialist approach to the subject of extrasensory perception, clairvoyance, and telekinesis began roughly at the same time as the famous "Airship Sightings" of the end of the nineteenth century (a subject we will get to in the next volume in this series). There had been stories of possible life on other planets for centuries, but the arrival of the Airships did not always reflect that idea. There was speculation that perhaps these mysterious sky ships were a new invention, something designed in a secret laboratory or a workshop far from prying eyes, rather than visitors from space. At the same time, and running in parallel, there was renewed interest in spiritualism, contact with the dead, and occultism (an interest that never really died, even in America, but which underwent cycles of interest from time to time). There was no overt connection seen between the ships in the sky and paranormal experiences, however; that would not begin to take place until the twentieth century was well under way, by which time science had put everything that was not reducible to rational explanation into the same basket, thus linking the paranormal with Ufology and mysticism in a kind of "guilt by association."

Actually, the scientists were not far off in their reasoning.

▼　　▼　　▼

The impulse to bring paranormal experiences into the laboratory for scientific testing and appraisal began in the late nineteenth century, when skeptics started investigating the claims of spiritualist mediums. The Society for Psychical Research (SPR) was founded in London in 1882, and boasted members such as William James (author of the enormously influential text *The Varieties of Religious Experience*) and Henri Bergson, among many other scientists and philosophers, including several Nobel laureates. The London SPR was followed three years later by the American Society for Psychical Research (ASPR), largely due to the influence of William James. The mission of both groups was to investigate telepathy and hypnotism, as well as spiritualism, séances, and the like, from a scientific perspective.

This was the same period that saw the creation of the Theosophical Society (1875), an organization based on the writings of Helena Blavatsky, a medium who believed herself (or portrayed herself) to be in contact with various incarnate and disincarnate entities who were giving her secret teachings. Her brand of theosophy was a reaction to Darwinism, which was itself seen as an assault on organized religion and especially the Abrahamic religions. By incorporating some Darwinian ideas—such as evolution—into a spiritual framework, she offered a path whereby one could appreciate the advances of science while maintaining a belief in human spirituality. Her two major works—*Isis Unveiled* (1877) and *The Secret Doctrine* (1888)—remain influential texts in alternative religion circles with their emphasis on creating a bridge between Western mysticism and the religions of Asia, primarily India.

This was also the era of European adventurers traveling to Central Asia and Tibet. Among those were Sven Hedin, L. A. Waddell, and, later, Alexandra David-Neel as well as Blavatsky herself. There was a romantic, Orientalist mystique associated with that part of the world along with the suspicion that the shamans and priests there had access to secret techniques of expanding consciousness, controlling the forces of nature, and communicating with the invisible world. One hundred years later those same concerns would form the basis of modern scientific inquiry, especially in connection with quantum consciousness and artificial intelligence.

It is well known by now that Harry Houdini joined forces with Sir Arthur Conan Doyle—the author of the Sherlock Holmes stories—in order to catch fraudulent mediums in the act, but also to identify those who might have a genuine gift. Houdini, as a stage magician, believed all mediumship was false, and was thus the philosophical ancestor of today's Amazing Randi. Doyle was not so sure; he had had paranormal experiences of his own and was willing to be convinced that genuine mediumship did exist. Their collaboration began after World War I, when Doyle was certain he had received postmortem communication from his son, who had died in that war. The two men jointly investigated claims of the paranormal, but parted ways after one séance in which Doyle claimed that Houdini had heard from his dead mother, an assertion Houdini heatedly denied.

By the 1930s, there was a more intense effort by some American scientists to codify a discipline that would once and for all identify and quantify psychic phenomena. It was suggested that a factor existed in the human mind—or perhaps in the world at large—called Psi, which was responsible for paranormal abilities. The Greek letter *Psi* is the first letter in the word *psyche*, or "soul," and was used to denote this mysterious force. At that time, the phrase "psychic phenomena" was being replaced by the more austere term *parapsychology*. During this same time, Joseph and Louisa Rhine set up an ESP laboratory at Duke University, which would become famous as the center for scientific testing of extrasensory perception. It has remained so to this day.[115]

It wouldn't be until the 1970s, however, that a concerted effort was undertaken by American intelligence and military organizations to get to the bottom of ESP, telekinesis, and associated phenomena, following the realization that the Russians and the Chinese both were aggressively studying it for military applications. The theoretical and practical advancements made in cybernetics, artificial intelligence, computer science, and robotics since the post-war period had contributed to a sense that the missing piece in all the equations might be what the people at Rhine were calling Psi. The laboratory experiments with student volunteers at Duke had presented them with mixed results; it seemed reasonable to suppose that a military or intelligence program could succeed where the civilian efforts had not. And if the Russians and the Chinese were working on the problem—and these were officially "atheist" countries that did not believe in the religious or occult aspects of the field—there had to be something to it. It had to be a science, a *real* science, and not some New Age pseudo-science, if the KGB was invested. At any rate, it would not do to have scientists in the Communist Bloc discover the operating principles behind paranormal abilities first, for the result could be as paradigm-shifting as the atomic bomb. And as lethal.

The world had already seen the weird power the Chinese had over American prisoners of war during the Korean conflict. Captured soldiers would be shown on television denouncing the United States and praising the Chinese and Korean regimes. The world learned a new term in 1950: "brainwashing." From the fear that there was some kind of "inscrutable Asian power to cloud men's minds"—born of incipient racism and "orientalism" tinged with genuine observation—the American mind control programs were born, both within the CIA and in the military. Hypnosis was one of the tools of the Bluebird, Artichoke, and MKULTRA programs, as was the use of drugs, especially hallucinogens. But the paranormal was being investigated as well, on the off-chance that mystics and shamans had stumbled upon some technology of the human psyche that unleashed latent mental or psychic abilities.

And it *was* an "off-chance." Generally speaking, the scientific community did not believe that there was any possibility that the decades of tests and experiments conducted by parapsychologists would yield real, repeatable, verifiable results. They pointed to poor testing protocols, the ever-present possibility of fraud or cheating, and reports that showed very little deviation from chance. Stories of paranormal events were treated as anecdotal and unrepeatable (much the same way UFO sightings were treated). Various psychological explanations were offered as a way to sideline any serious study of the matter.

Part of the problem was the way in which ESP and other tests were conceived. It was felt that if a person was considered "psychic," that ability should be testable using Zener cards[116] (Figure 16) or some other type of symbol set or technique that could be manipulated under laboratory conditions. Psychics who refused to submit to such a bizarre regimen would be considered charlatans who were afraid of being exposed. If they did submit, but after hours or days of constant, repetitive tests they did see an opportunity to cheat and took it out of frustration, they naturally were accused of being charlatans.

Figure 16. The Zener cards: a set of five symbols used to test Psi.

That there may be an emotional and situational aspect to telepathy, psychokinesis, or even remote viewing, generally was ignored because those are components that cannot be factored into a laboratory test. There have been many well-documented, if necessarily anecdotal, cases of precognition or extrasensory perception involving family members who sense emergencies taking place with their loved ones, including traffic accidents, violence, etc. This type of experience cannot be adjusted for in the laboratory unless one were to kidnap and subject a family member to horrific circumstances in order to see if the test subject reacted. Even then, due to the contrived nature of the kidnapping and abuse, one would not be duplicating the original experience.

Thus, the emotional component is possibly necessary to the average experience of ESP which is, after all, a consciousness artifact; the contextual element is another one that is never factored into the experimentation.

The tests for ESP, PK (psychokinesis or telekinesis), and other paranormal abilities usually take place in a sterile, controlled environment, which is not the normal environmental context for most individuals. It immediately raises a degree of tension and stress that some people associate with being in a classroom or a doctor's office. Why should any of that matter, though?

If, as some scientists have suggested, consciousness is an "emergent property of the brain," then ESP, PK, and other paranormal abilities also may be emergent properties. The ESP test is not an intelligence test or a test of normal human perception, so using the same format and same mechanistic approach may not be suitable. After all, the type of perception being tested is deliberately named "*extra*sensory," implying a type of perception that *may be* located in the brain but does not utilize the "macro" physical senses themselves, neither for receiving nor sending information.

Thus, the conscious state of the test subject—as complex a state as it is for anyone, even under ordinary circumstances—must be taken into account when it comes to testing for ESP. Further, the conscious state of the test subject is modified or enhanced by what the LSD experimenters of the 1960s called "set and setting," and what shamans and ritual magicians for millennia have called the ritual space. This extreme approach, equivalent to the sensory deprivation or "sensory invariance" experiments of the fledgling US space program that we saw in the previous chapter, is capable of causing hallucinations, but in some cultures it is also a necessary platform for the development of paranormal abilities. As far as we know, this type of approach has not been attempted by laboratory parapsychologists, as it would smack of superstition and occultism, which paradoxically are in the same neighborhood as the paranormal abilities they want to test in the first place. Thus, science in general has been successful in resisting or denying the value of this type of research, as it challenges their entire paradigm.

In addition, the classical model of physics does not permit the existence of psychic phenomena. For instance, if there are only four forces at work in the universe—electromagnetism, the strong nuclear force, the weak nuclear force, and gravity—which one would be responsible for my moving an object using only my mind? How much energy would need to be expended, in violation of Newton's laws? Merely postulating the existence of Psi does not prove it. It remained elusive, even in J. B. Rhine's fancy labs at Duke. Eventually the university severed its relationship with the center, and it was forced to operate autonomously.

At the same time, however, the US military and intelligence agencies were well aware of ongoing research into Psi by their ideological enemies. To ignore the Soviet and Chinese programs was tempting—"Let them waste their time and money with this nonsense"—but knowing that

Russia beat the United States into space because they were willing to entertain a multidisciplinary approach to the problem, it was incumbent upon the American national security apparatus to at least consider the possibility that there might be something to it.

Enter the Stanford Research Institute, later to be known as SRI International or simply SRI.

Founded in 1946 as part of Stanford University, SRI was already independent and functioning as a nonprofit research institute by the time the US government called upon its services in 1970. Its first contracts were for the US Department of Agriculture and the newly created United States Air Force. Other clients included Walt Disney, Bank of America, the Technicolor Corporation, and many other household names. SRI became involved in the field of artificial intelligence and robotics, designed the first computer mouse, and of course was where ARPANET—the forerunner of the Internet—was developed.

But the Vietnam War changed the trajectory of SRI. In 1970, due to pressure from antiwar groups who saw SRI as part of the military-industrial complex, it was divested from Stanford University. Two years later, an extraordinary project with national security implications was begun.

▼   ▼   ▼

Hal Puthoff and Russell Targ were laser physicists who had worked at different Sperry Gyroscope locations before meeting each other in California. When they became interested in Psi as an outgrowth of their research into quantum biology (and encouraged by CIA) they made the acquaintance of the artist Ingo Swann, who was an accomplished psychic from New York. Eventually they also became acquainted with retired police officer Pat Price, intelligence officer Joseph McMoneagle, and perhaps most famously with Israeli psychic Uri Geller. Puthoff himself previously had been a naval intelligence officer and had been employed as a civilian for the National Security Agency.

Swann had come across a proposal by Puthoff for research involving the measurement of plants and "lower organisms" to determine "whether physical theory as we knew it was capable of describing life processes."[117] When Puthoff was in New York visiting his friend Cleve Backster, who was involved with attaching polygraph equipment to plants to measure their reactions to stimuli and who had a copy of the proposal, Swann contacted him to ask if he had ever considered investigating parapsychology as part of his research in physical theory and quantum biology. Swann claimed to have performed PK experiments successfully in a lab at City College. Intrigued, Puthoff invited him to the Menlo Park offices of SRI for some testing.

Puthoff had set up an experiment in which Swann would try to "perturb" a heavily shielded magnetometer at Stanford University from several stories above. Swann not only affected the device but went on to "remote view" and then sketch the complicated equipment in some detail, even though its physical features had never been published in any form. This led to a visit by the CIA to Puthoff's office and further testing of Swann, which resulted in the agency giving SRI an eight-month contract worth nearly fifty thousand dollars. Russell Targ joined the team, and the remote viewing program was born.

Ingo Swann's capabilities earned him considerable government attention. At one point he even psychically explored the moon, the planet Jupiter, and the UFO Phenomenon, as he recounts in his book *Penetration*. In his remote viewing of Jupiter, for instance, he revealed that the planet had a ring around it, similar to Saturn. This was dismissed by astronomers out of hand, until the NASA Pioneer 10 space probe flew past Jupiter and confirmed Swann's report.

This was only one example of someone working in the field of ESP and PK becoming involved in Ufology as well. Uri Geller—whose most celebrated specialty was bending spoons using only his mental abilities—was a protégée of inventor and doctor Andrija Puharich, who believed he was in contact with an extraterrestrial source—called Spectra—and who then managed to convince Geller of the same, at least for a while. Puharich's extraterrestrial contacts went back to 1952 and 1953, when séances he conducted at his home in Maine formed a link with a group called "The Nine," which, as it turned out, was another name for Spectra. Spectra, as we will see in greater detail in Book Three, was said to be an intelligence on a spacecraft hovering in near-Earth orbit that was operating through select human beings on the planet to inspire or actually to cause radical change.

Puharich's involvement with SRI and the ongoing Psi testing by Puthoff and Targ is beyond the scope of this chapter, and in any event he was only one of many individuals—famous and not-so-famous—who filed through the Menlo Park labs. What is of interest is the growing interest by the national security apparatus in what was happening at SRI and how Psi could be exploited for intelligence purposes.

Thus was born a series of special access government programs designed to test and expand Psi capabilities—basically, to weaponize Psi—that went on for years under various agency controls, from the military to the Defense Intelligence Agency (DIA) and the CIA. Psi research needs funding—for laboratory space, equipment, salaries, and travel—and there are few potential sources for this type of funding. One would be looking either at the military or the intelligence services; the private sector Psi labs were often defense subcontractors, like SRI. The intelligence services themselves were CIA and DIA, as well as the Air Force, the US Army's INSCOM,[118] and the Department of the Navy.

Due to the declassification of thousands of pages of documents in 1995, 2001, and later concerning these government programs, we now know a great deal about how SRI and other institutions studied telepathy and other parapsychological effects. While some of the documentation is predictably dry and full of bureaucratic jargon, other pages reveal much about the testing itself and the uses to which the government-sponsored psychics were applied, such as in the search for missing politicians or spying on remote military installations.

Ingo Swann was the first psychic to be tested under this program, and he would come to earn a "top secret" security clearance in the process.[119] He came to the attention of the world at the New York City headquarters of the American Society for Psychical Research (ASPR) in 1972, where he had an exhibit of his artwork during a reception on April 26 of that year. It was reported that "Mr. Swann is participating extensively in ASPR experimentation on out-of-body states."[120] It was that same year, in June, that Swann began his relationship with SRI.

By 1973, Puthoff and Targ were ready to issue a report on the work they had undertaken with both Ingo Swann and Uri Geller. A declassified memo in CIA archives reads, in part:

> Dr. H. Puthoff and Mr. R. Targ will present their final report of parapsychology studies completed at Stanford Research Institute (SRI) at 1000 hours, 10 September 1973, in the East Building conference room. The report will include but not be limited to results obtained from studies of two exceptionally capable individuals, Mr. Uri Geller and Mr. Ingo Swann. Experimental procedures, data and conclusions will be presented.[121]

It's important to realize that Swann was instrumental in validating the Psi project at SRI, since his results were pretty consistent. He performed some astonishing feats of telepathy and what eventually would become known as "remote viewing." The military and intelligence services wanted to know how they could exploit this type of ability on a grander scale. Could anyone be trained to use Psi? What were the psychological and physical requirements? How much training—and what kind—was necessary? How fast could a team of remote viewers with security clearances get up and running?

The problem was that science had not found a way to describe how Psi worked, much less predict it with any kind of certainty. While researchers worked on developing some palatable scientific explanation for Psi that would satisfy the more cynical of their clients, they also had to develop systems for training and applying the paranormal powers. Ordinarily, one cannot do that without first understanding what one is working with, and this they did not have.

In the 1970s, the type of consciousness studies they would need as theoretical scaffolding for their work did not yet exist. There was a lot of buzz concerning quantum mechanics (QM) and its use as a paradigm for explaining (and hopefully re-creating) Psi abilities, but it was even more hypothetical then than it is today.

Enter Arthur Koestler.

The Scientists and the Mystics

As a final touch of controversy, I might add that the existence of lawlike time reversal, especially in the form of negentropy—information reversibility, together with time extendedness in relativity theory and arrowless causality in the EPR correlations, have convinced me that there might very well be some truth in 'the claims of the paranormal.'[122]

—Olivier Costa de Beauregard,
Time: the Physical Magnitude

Puthoff and Targ presented two papers at a conference in Geneva, Switzerland, in August of 1974, which was the year after the report on Geller and Swann mentioned in the previous section. The conference theme was "Quantum Physics and Parapsychology" and was under the aegis of the 23rd Annual International Conference of the Parapsychology Foundation. A report on the conference dated October 28, 1974, was included among the declassified CIA files on the subject, but the name of its author has been redacted. That the CIA would have sent someone to observe and report on a parapsychology conference should come as no surprise, and it is of interest to note that the link between quantum physics and parapsychology was the subject of an entire conference as early as 1974.

The conference theme was the inspiration of Arthur Koestler, the well-known novelist and philosopher whose book *The Roots of Coincidence* was a bestseller at the time. Indeed, coincidence—or, as the Jungians call it, synchronicity—had been a fascination of some physicists for a while, including the Nobel Prize–winning scientist

Wolfgang Pauli, whose correspondence with the Swiss psychiatrist Carl G. Jung on synchronicity has become part of the canon.[123] Other scientists, such as F. David Peat and David Bohm, have contributed to discussions on the subject over the years.[124] Jung and Pauli considered synchronicity to be an "acausal connecting principle," which meant that events were connected in a way that defied our normal cause-and-effect understanding of the physical world. Synchronicity therefore seemed to have a lot in common with some of the core concepts of quantum mechanics, such as entanglement and non-locality. The idea that there could be a scientific explanation for Psi—even if it was a quantum explanation and not something out of Newton's or even Einstein's playbook—was seductive, and synchronicity (embraced after all by Nobel Prize laureate Pauli) seemed to point in that direction.

The CIA report by the anonymous author summed up the proceedings succinctly:

> The reality of ESP was an accepted fact at this conference. When discussion of the reality of the phenomena occurred, it was concerned only with methods of securing public acceptance and belief in paranormal cognition. . . . It became clear that there exists at present *no* adequate theory of paranormal perception which can furnish a physical basis for the phenomena. The existing attempts at physical theories are speculative, incomplete, at

best poorly substantiated; although some ideas were presented which might furnish the seeds of fruitful investigation.[125]

And further:

The conference evidenced a rough division of attitudes toward parapsychology into two schools of thought. Researchers share a general acceptance of the strangeness of it all, but on the working level some believe progress in understanding can be achieved through extension and use of existing scientific knowledge and methods, perhaps by some new synthesis. The others basically favor a more philosophical and mystical approach, and believe that nothing short of a complete revolution of thought, maybe into more spiritual directions, can cope with this challenge.[126]

It is interesting to compare that observation with the sentiment of Ingo Swann, expressed at the beginning of this chapter, which would place Swann in the "mystics" category. However, Puthoff and Targ represented the scientific point of view, which of course they demonstrated in the two papers they presented. The first of these was titled "Physics, Entropy and Psychokinesis" and the second "Remote Viewing of Natural Targets." The author of the observational report commented, regarding the Puthoff/Targ presentations:

Belief in the experimental results depends upon confidence in the integrity and skill of the experimenters. The results described, if valid, establish the existence of the phenomena and are a first step toward establishing its patterns.[127]

This was one of the more positive reviews of the mysterious author, who saved his more scathing attacks for the "mystics." He reserved his highest praise, however, for the eminent French physicist and philosopher of science Olivier Costa de Beauregard, who presented a paper titled "Quantum Paradoxes and Aristotle's Twofold Information Concept." One of Costa de Beauregard's primary interests was in what he called "physical time" and "time reversibility" (or the lack of it: time as an arrow, which proceeds from past to future and is linked to cause and effect). Costa de Beauregard suggested that information theory was a key element in any discussion of ESP. As the CIA report states:

It provides a theoretical framework which not only may connect observed ESP effects with quantum mechanics, but which also places ESP in the very general context of information theory. Whatever else it may be or may involve, ESP does deal with information. . . . This is a provocative paper, perhaps the most profound physical discussion at the conference.[128]

The author went on to connect the papers of Puthoff and Targ with the presentation by Costa de Beauregard in an approving way, which was high praise coming from him.

We've spent some time on this conference and the viewpoint of a CIA observer for a reason. It was 1974, and official US government interest in the paranormal was just ramping up to speed. The initial experiments with Ingo Swann and Uri Geller would become the basis for a full-scale investigation of the paranormal by the US Army, the DIA, and the CIA that would involve many more test subjects and missions that were increasingly sensitive. These special access programs would be known by a variety of code names as they progressed from one iteration to the next, one "client" to the next, one funding mechanism to the next, over a course of more than twenty-five years. We would come across SCANATE (1970), GONDOLA WISH (1977), GRILL FLAME (1978), CENTER LANE (1983), SUN STREAK (1985), and, of course, STAR GATE (1991), all versions of the original project as undertaken by Hal Puthoff and Russell Targ. A full discussion is beyond our scope, but the declassified files that detail their efforts are a window into how the scientific establishment was studying, measuring, and testing Psi in all its forms: ESP, PK, precognition, and of course remote viewing.

A Defense Intelligence Agency (DIA) report—which is undated but probably was created in 1994—provides an excellent introduction to how the defense establishment defined and tasked Psi:

DEFINITIONS

PSYCHOENERGETICS

- A mental process by which an Individual Perceives, Communicates with, and/or Perturbs Characteristics of a Designated Target, Person, or Event Remote in Space and/or Time from that Individual.

PSYCHOKINESIS

- Physical Actions Performed by Mental Powers that cannot be Explained by known Physical Means.

ESP & TELEPATHY

- Perceptions which cannot be explained by Known Sensory Means.

REMOTE VIEWING

- The Acquisition and Description, by Mental Means, of Information Blocked from Ordinary Perception by Distance, Shield, or Time.

REMOTE VIEWER

- A Person who Perceives, Communicates with, and/or Perturbs Characteristics of a Designed Target, Person, or Event.[129]

And this is only a taste of the rest of the document, which also mentions "Extended Remote Viewing" (ERV), which is designed to achieve "an Altered View of Reality"; "Coordinate Remote Viewing" (CRV); and "Written Remote Viewing" (WRV). In addition, "Secondary Methodologies" include Dowsing, Psychometry, and Clairvoyance. Sample test reports were developed, as well as questionnaires, and other requirements as needed. (It is frankly mind-blowing to read about "clairvoyance" in a DIA report, even now.)

A battery of psychological tests was administered to those who volunteered for these experiments, including the Minnesota Multiphasic Personality Inventory, the Gordon Personal Profile Inventory, the Fundamental Interpersonal Relations Orientation–Behavior (FIRO–B), the California Psychological Inventory, the Edwards Personal Preference Schedule, and the Personal Orientation Summary. This was in addition to eight blood tests, waking EEGs, sleeping EEGs, and numerous other examinations.

A certain class of test subjects was automatically rejected:

> Individuals who displayed an unreasonable enthusiasm for psychoenergetics, "occult fanatics," and "mystical zealots" were not considered for final selection.[130]

And yet:

> It seemed that the best psychics would be more likely to exhibit EEG "asynchronicities" or patterns of abnormally imbalanced electrical activity,

between the two halves of the brain. One often found such EEG asynchronicities in individuals with minor seizure disorders or hyperactivity disorders. . . . Some of them had had relatively intense dissociative experiences, had high IQs, and clearly didn't fit into the world the same way other people did. And yet for the most part they were functional, emotionally healthy people—in some cases, much more functional and healthy than the average person.[131]

The Shaman and the Psychopath

Like the sick man, the religious man is projected onto a vital plane that shows him the fundamental data of human existence, that is, solitude, danger, hostility of the surrounding world. But the primitive magician, the medicine man, or the shaman is not only a sick man, he is, above all, a sick man who has been cured, who has succeeded in curing himself. . . . The initiation of the candidate is equivalent to a cure.[132]

—Mircea Eliade

There was considerable concern that psychologically unstable persons would be attracted to this project, and they were presumed to include those who were actual occultists and mystics. In fact, at least one scientific paper was published in the literature that suggested there is a link between mental illness and paranormal abilities.

"Research into Psychotic Symptoms: Are There Implications for Parapsychologists?" bears the following statement in its abstract:

> Both parapsychology and psychopathology deal with anomalous experiences. Moreover, statistical associations have been reported between paranormal experiences and psychological symptoms.[133]

This is a problematic stance to take, since the psychologically unstable in many cultures are precisely those persons who demonstrate "occult" powers. The paper draws a link between the experience of hallucinations in those deemed mentally unstable, for instance, and mystics or occultists who claim to have visions. However, it was pointed out that "psychiatric disorders exist on continua with normal functioning, making any clear division between the 'them' who are mentally ill and the 'us' who are untroubled difficult and in many ways unhelpful."[134]

The author of the paper cites a 1991 Gallup poll that found "a quarter of those surveyed believed in ghosts . . . and about one in seven thought that they had seen a UFO," thus linking UFO experiencers with those who believe in ghosts; in other words, linking what is believed to be an irrational belief in ghosts to those who actually report seeing a UFO. As we mentioned earlier, this is the tendency to put all such "unscientific" ideas into the same basket.

The handwriting was on the wall: until recently, expressing a belief in the Phenomenon meant that one suffered from delusions and mental illness.

> Some studies indicate that people who report paranormal experiences have higher than normal levels of psychological symptoms . . . whereas others indicate that individuals suffering from mental illness report unusually strong convictions about the reality of supernatural forces.[135]

John Mack, the famous Harvard psychiatrist who committed the mortal sin of taking abductee experiences seriously, agreed that this was a problem but approached it differently:

> The phenomenon does not stand alone, but is one anomaly among many. . . . These include evidence for clairvoyance, telepathic communication, remote viewing, psychokinesis, non-locality, the demonstrated efficacy of prayer. . . . They reveal that our understanding of reality is extremely limited, the cosmos is more mysterious than we have imagined, there are other intelligences all about.[136]

That UFO experiencers and particularly abductees may have been suffering from a mental or psychological disorder is what prompted Mack to begin studying them in the first place. It soon became apparent that the "disorder" had to have an actual cause that could not be identified as an organic neurological imbalance or due to drugs, hypnosis, hysteria, etc. In other words, one may appear to be suffering from a mental disorder but in reality is reacting to a trauma that actually occurred. It is virtually impossible for the average modern human being to tell the difference between someone who is mentally ill and one who has had an experience so traumatic that it has reordered their worldview.

The study of the intersection between mental illness and mental telepathy is what gave rise to this discovery. Was a belief in mental telepathy evidence of psychological problems, something that could be cured with antipsychotic medication and intensive therapy? Was belief that one had been abducted by aliens the same type of mental dysfunction?

If all we had to work with was the science of Isaac Newton and Albert Einstein, perhaps. The mechanistic, reductionist approach would be consistent with the scientific worldview. But anthropologists and historians of religion have been telling us for decades that the answer is not so cut-and-dried. In cultures where science and technology have been late to the party, there is greater acceptance of persons we would diagnose as mentally ill, even schizophrenic, but usually under certain very specific circumstances. There are shamans who go into the wilderness for days, weeks, or months at a time and experience every type of horrific scene of torture and death, only to return to their villages empowered to heal, to divine the future, to see the invisible world. Are there not parallels between the shaman who is cured—to use Eliade's phrasing—and the abductee who returns to her home after the hideous experience of being surgically operated on but who then has visions of an impending apocalypse?

Of course, there are cultural differences between the Siberian shaman and an abductee in New Hampshire or Arizona that it would not be wise to ignore. There are also such differences between the experiences of white Americans, for instance, and African-Americans, for whom the motif of the strange "night doctors" who come in the middle of the night to kidnap them for surgical purposes reflect fears about a reality only all too tangible.[137] It has been suggested that the "night doctors" story was invented by "southern whites anxious to prevent their labor force from leaving the region."[138] If so, is it possible that the UFO abductee experience had a similar genesis, perhaps among "deep state" actors eager to inspire fear of the UFO in order to justify their own space-based military programs and expenditures?

The difference is that the UFO abductees *actually experience* the abductions and resultant interference with their bodies and their lives. The "night doctors" story was part of a rumor stream, and no one actually came back from a session with the night doctors to tell the tale. The night doctor story had its origins in the very real atmosphere of fear and intimidation associated with white segregationist groups in the South, such as the Ku Klux Klan, which were carrying out lynchings and murder against the African-American population. In addition, there was the fact of medical experimentation projects such as the Tuskegee syphilis program, which targeted the African-American community and lasted well into the 1970s. So the idea of evil alien doctors coming in the middle of the night to abduct and operate on innocent African-American citizens was credible to a certain degree; the alien abduction scenario, however, was considered less of a "cultural artifact" or the expression of a political reality than it was evidence of a mental disorder.

That all began to change with the work of John Mack, Budd Hopkins, and others who took the claims of the abductees seriously. Once that bridge was crossed, it became easier to credit stories of paranormal experiences that accompanied (or were the result of) the abductions, such as the many cases of mental telepathy that were reported. If the abductions were real *in some way*, then maybe there was a basis for believing in the possibility of ESP, PK, and the rest. This was about consciousness, and information, and the Other. Ingo Swann was seeing UFOs *and* aliens, and had a security clearance, but he was also remote viewing for the government, and there was never any indication that *he* was mentally disturbed. In spite of this, until the application of quantum mechanics to consciousness studies, there simply was no scientific theory that would permit the existence of these powers, just as there was no scientific theory that permitted the reality of the UFO or the abductions.

▼　　　▼　　　▼

Quantum mechanics would provide the inspiration for a generation of scientists-cum-mystics who saw the potential for a scientific explanation of Psi that would take ESP and the paranormal out of the closet and into the light of day, and thereby rescue some of those who claimed to have telepathic abilities from the ministrations of the therapist, psychiatrist, and mental health clinic. That meant, however, that the UFO Phenomenon was also in danger of being "revealed" by the new technologies and the new philosophical approach. Hal Puthoff of SRI was a physicist who was interested in UFOs as well as in the possibilities of remote viewing.[139] Jung, the psychologist with decades of experience in psychotherapy and depth analysis, believed UFOs were nothing more than the projections of unconscious anxieties; a sentiment he shared with Wolfgang Pauli, publishing a monograph on the subject.[140] Another Pauli correspondent, the scientist Max Knoll of Princeton and the University of Munich, was unequivocal in his dismissal of the UFO Phenomenon:

> I would like to say that once again Jung must be made to understand that the UFOs seen on radar screens are no more 'real' than those sighted directly, and that no definite conclusions can be drawn about their actual existence except by radar photographs or radar films (examined by experts). . . . Based on the observations that have been made so far, Jung would actually be in a position 'to give a convincing denial of the objective existence of UFOs.'[141]

It is interesting to note Knoll's insistence that Jung make a public statement denying the reality of the UFO Phenomenon. Clearly, Knoll wants Jung's psychological explanation to be the dominant one. Knoll was the co-inventor of the electron microscope, along with his student Ernst Ruska, who would go on to win a Nobel Prize. Both Knoll and Ruska remained in Nazi Germany during the war, continuing their research with a grant from Siemens. At the time he wrote the letter to Pauli, Knoll was studying brainwaves and the electrical stimulation of the brain.

The passion that the study of telepathy and associated paranormal phenomena arouses in both the "scientist" and the "mystic"—to use the CIA's terminology—reflects a sea change taking place in the culture. It signals the dawning of a new realization about the world we live in. Science wants to make it simple, to make it predictable and mechanical. Life as machine. But the existence of the UFO Phenomenon means that the battle has not yet been won; that there is a "paranormal machine" that has to be explained and disassembled.

The individuals at the CIA and DIA and in the military who were most relied upon to use their paranormal training for national security were the ones who entertained awkward views concerning reality, parapsychology, and alien existence. If life as we know it is no more than a machine, then there is a "ghost" in the machine, as Arthur Koestler would say.[142]

The idea that the mind is an emergent property of the brain, which does not exist independently of the body, was central to Koestler's thesis. That did not mean that he discounted Psi or parapsychology in general, but that he felt it had to be an emergent property of the brain as well. In other words, the mind was not a "ghost in the machine" of the body, but a property of the body.

This is the impulse behind much development in artificial intelligence, of course. If a machine is built that is powerful enough, with a brain the equal in computing power to that of a human, it will develop consciousness. It should also, then, develop paranormal abilities. It should develop Psi.

If it doesn't, then perhaps the mind is not an artifact of the body.

If it does, then even if the "aliens" are no more than machines themselves, they may exhibit not only consciousness but also paranormal abilities; perhaps ESP, PK, and precognition, but also abilities we have yet to imagine (or to take seriously).

If, as Costa de Beauregard suggested, time is reversible under some circumstances (particularly in the field of information theory), then the telepathic communications that abductees have with their alien abductors may be an indication of greater abilities, such as an ability to time travel.

The telepathic communication of information experienced by abductees may be the *message* itself, *not the content* of the images and communications. (The medium *is* the message, as Marshall McLuhan[143] said.) Their effortless use of telepathy in their interactions with humans may signify much more than a simple way to get around the fact of language. With telepathy, is anything truly "lost in translation"?

As Ingo Swann, government telepath and UFO experiencer, writes:

> Most surprisingly, one might think that Ufologists would consider mental processes of extraterrestrials, since they are so energetically involved with extraterrestrial equipment and technology.
>
> None of the above will touch the topic of Psi with a ten-foot pole, and all of the above protest any feasible, positive necessity for acting any other way—although some psychologists studying abduction phenomena have begun to notice the telepathic factor.[144]

Imagine having a television set you could never turn off; worse, imagine it inside your head, and someone else could switch it on or off at will. Researchers at MIT[145] have already developed a device that allows you to control your computer with thoughts alone. The goal is to enable a user to communicate with their smartphone without actually looking at the phone or interacting with it physically. It's only a short step from there to enabling a user to communicate with *your* smartphone and from there to your own brain.

Suddenly, telepathic aliens don't seem so incredible.

Before we leave you with these somewhat unsettling revelations, there is one more to be considered.

As published in *Nature Neuroscience* in February 2018,[146] an international team of researchers has discovered that people who have had traumatic experiences—such as those suffering from PTSD—had their brains *rewired.*

In response to a traumatic experience, areas of the brain grow new connections involving a type of neuron called a "bridging neuron," which links the hippocampus to a region of the prefrontal cortex called the infralimbic cortex. This means that the brain rewires itself, forming new neural connections where none previously existed. The amygdala (the "fight or flight" center of the brain) actually *grows* in response to the trauma, but at the same time an entirely new set of connections is made in the brain that is *context-dependent* and which regulates memory retrieval.

This may be one reason why it is so difficult to treat patients with PTSD. Historically, therapists have tried either to focus on the traumatic memory in order to enable the sufferer to cope with it directly, or to submerge the memory using a variety of therapies and drugs. The problem with these methods is that the actual memory of the trauma can resurface at any time.

Now researchers think they know why. The brain restructures itself in response to the trauma, creating a neural system that accommodates that trauma as part of its overall design. While a person undergoing therapy in a clinical setting may experience temporary relief from the traumatic memory, once outside the clinic the memory can resurface. That indicates a context-dependent mechanism: the association of a specific context with a specific memory.

If we apply this new research to the alien abductee, we can begin to understand why some individuals claim they have been modified in some way by the experience, that their brains have been changed. It may even help to explain the implant experience. The alien abduction experience is traumatic; John Mack described the abductees as suffering from a form of PTSD, which indicated that they had indeed suffered a genuine traumatic experience. The unusual context of their experience is such that it incorporates not only the trauma of sexual abuse and of kidnapping—common to many PTSD sufferers who have experienced these events in real life—but also the trauma of confrontation and direct contact with nonhuman entities, along with a host of paranormal events (such as levitation, telepathy, etc.). If the normal hideousness that takes place in normal, everyday life (violence, rape, abuse) can rewire the brain in response, how much more so the experience of alien abduction no matter what one may think of its reality?

> In reality I do know something about UFOs now . . . I also think the phenomenon is physical, but it integrates psychic functioning. Contact is possible but irreversible.[147]

This would apply not only to the UFOs themselves, but to their purported occupants, the Aliens or Visitors. If the UFO can seemingly violate Newtonian physics in its aeronautical maneuvers, air speed, and ability to predict where an F-18 Hornet will be in the future (as former AATIP director Luis Elizondo has stated in several interviews), this ability to manipulate matter where machinery is concerned might very well extend to the way in which the Visitors seem able to move human beings through walls, levitate them, and perform other, similar actions that are normally beyond human capability. It may all be the same science: that the UFO moving through space and the abductee moving through space are aspects of an identical process.

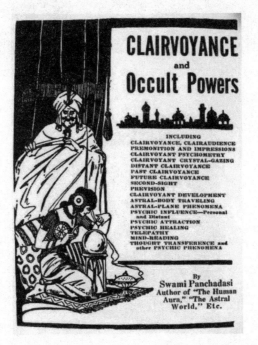

A pulpy pamphlet on occultism, but nevertheless a pretty good list of paranormal abilities that were tested by the US Government through such subcontractors as SRI.

J. B. Rhine of Duke University (right), testing a subject for Psi abilities.

Dr. Karlis Osis of the American Society for Psychical Research in Manhattan, testing a subject for Psi.

A photograph of a man "levitating" a chair: an example of PK (psychokinesis).

Psychic Stanislawa Tomczyk levitating a pair of scissors, 1909.

GRAY ANATOMY

Perhaps *homo sapiens*, the wise, is himself only a mechanism in a parasitic cycle, an instrument for the transference, ultimately, of a more invulnerable and heartless vision of himself.

– Loren Eiseley[148]

ALL THE FOREGOING SHOULD HAVE SUGGESTED TO YOU that there is ample reason to draw some conclusions about the Phenomenon and how it interacts with human beings. We have seen how consciousness studies struggle with these same issues, but in a different—anthropocentric—way, and if we could shift the focus from humans to nonhumans we may arrive at some useful (if startling) conclusions. We have seen evidence that suggests that even our genetic code is an alien inheritance and that somehow our ancestors were aware of this and tried to represent the double helix and the code both graphically and mathematically, respectively.

We've also looked at the developing technologies of artificial intelligence, quantum mechanics, and even mental telepathy and how these relatively new disciplines are bringing us closer to explaining the experiences of alien abductees and contactees as our own science approaches the characteristics of what has been described of alien contact.

While all of this does not definitely answer the questions "What are they, and why are they here?," we will try in this chapter to get a little closer to actually seeing the Visitors for what they are.

For that, we have to remember everything we already have learned and apply it a bit more proactively to the central problem of the "alien" itself.

There have been a number of well-known specialists in this field, ranging from Budd Hopkins to John Mack, from C. D. B. Bryan to David Jacobs. Whitley Strieber has given us detailed descriptions of his own experiences, which serve to flesh out some of the other accounts, and Jeffrey Kripal has made a serious contribution to the field by collaborating with Strieber to provide a more nuanced yet still quite adventurous exploration of some of the more mystical, religious, and philosophical implications and explanations of alien-human contact. And of course Jacques Vallée has explored this field in considerable depth and has come away with the view that the UFO Phenomenon is much more than merely seeing lights in the sky, or aliens in a flying saucer.

Although these authors are often in conflict with each other over the social and political implications of contact, they nonetheless agree on some essential points, and we should start there. Our goal in this section is to come as close as possible to exposing the "Visitors" for what they really are—an ambitious target, to be sure, and one that assumes a great deal of information "not in evidence," as they say in courts of law. What evidence there is, however, should be sufficient to start this conversation going in the right direction. Unless you have been a contactee, experiencer, or abductee yourself, it is nearly impossible to understand the core qualities of the event(s), but we will try to form a basis for such understanding.

Alien Phenotypes—North America

Aliens come in different sizes, shapes, and overall appearances: their "phenotypes," to borrow a term from biology and genetics. The most common—the most reported upon, the most depicted in art and culture—is the iconic "Gray." This is a being usually shorter than average human height, perhaps no taller than four feet, or a little over one and two-tenths meters. They exhibit a vestigial nose and mouth, no ears, and no external genitalia. Their heads are abnormally large, with large (usually black) eyes. They seem to have no knees, and their gait is what we might call "robotic." They are gray in color, from head to foot, and some contactees have suggested this might be a kind of suit and not their actual skin, but there is debate over this point. They are thin, spindly, and have elongated fingers in some cases. They are bipedal, that is, they have two legs, plus two arms, a central torso, and a neck that supports their head, which seems to indicate the presence of a spinal column, brain stem, and brain of some kind. They have two eyes, as we do. So in some essential characteristics they appear quite similar to humans.

They are never seen to eat or drink, or to excrete any substance such as waste products or saliva, mucus, etc. If they do not eat or drink or excrete, we may be safe in assuming that they are closer to machines than to organisms.

They communicate with gestures and with a form of mental telepathy; i.e., the contactees "hear" the Grays' speech in their heads, but not as sounds that could be recorded by audio equipment. (As we noted in the previous chapter, at least one scientist is predicting "brain to brain communication" by the year 2050 as a function of strong AI.)

That is how they appear. Now, what is it that they *do*?

The Grays have been objects of fear and trauma for many individuals because they seem ubiquitous in cases of alien abduction. When they appear, a period of "high strangeness" follows, which can be the sense that one is being transported through walls and through space to a hidden laboratory somewhere—on a spaceship, or elsewhere—at which time weird biological experiments take place. The abductees have lost all sense of agency during this event; they are helpless, cannot resist, and are subject to humiliating processes which seem to mimic or actually represent human scientific or medical operations and procedures. In many cases, these procedures are focused on human sexual reproduction in some way.

Eventually, the abductee is returned to the place where the abduction took place and often has no immediate memory of the event beyond the unpleasant feeling that something strange or even evil had transpired. Memories of the event return, either naturally over the course of time, or recovered through hypnosis or some other therapeutic procedure. Often with that recovery comes the realization that this event has happened before, perhaps many times before.

The Grays, however, are not the only alien phenotype. There are others.

There is the "taller, older" Gray that is sometimes observed. This Gray seems to be in charge of the smaller, "younger" Grays. A leader, or a commanding officer of some kind.

The most compelling phenotype, however, is what is sometimes called the Nordic type. Taller, often blond and blue-eyed, this version seems to convey a different message entirely. They are believed to be benevolent and concerned about the fate of humanity. Accounts of Nordic aliens go back at least to the 1950s and the reports disseminated by George Adamski, one of the first famous contactees and a man many believe to have been a fraud. That there may be a racial conceit buried in this experience cannot be ignored, especially as the darker-skinned Grays are perceived to be abductors and sadistic experimenters.

The third major, though less common alien phenotype is the Reptilian, which is a quasi-human, quasi-reptile being. The experience of Reptilians has been hijacked by conspiracy theorists, among them David Icke, as representing a hidden cabal of humanoid beings who control terrestrial governments. Ideas about "lizard people" can be traced back to the fantasy authors Robert E. Howard and his friend H. P. Lovecraft, while ideas about sentient lizards, snakes, and dragons are everywhere in ancient cultures. The fact that dragons, for instance, are considered good in Asia and evil in Europe should be an indication that there is no one human response to the type, which itself implies cultural and psychological reactions that are not shared and can influence the emotional sense of what is experienced. Yet, while it doesn't take a PhD in psychology to make some inferences about these three types of aliens and what they represent in human terms, that might be putting the cart before the horse.

We would be making a serious mistake if we assumed that what Americans perceive as alien beings—the three phenotypes we outlined previously—are perceptions that are shared cross-culturally and among different ethnic groups. It is just as important for this study and any to come afterward that different experiences of alien contact are reviewed and analyzed for variations and similarities. If we proceed on the assumption that the Phenomenon is real (something we have insisted upon since the very beginning), then the marked differences in how the alien is perceived are just as important as any similarities.

Alien Phenotypes—India

Reports coming from India indicate that both the Tall Nordics and the shorter Grays have made appearances in South Asia. One report, dating to 1992 and posted on an Indian website, shows that the typical alien abduction phenomenon is not a purely North American experience. A primary school student in Maharashtra reports seeing a white light in the middle of the night, levitation to a kind of craft, being accompanied by "Four to five tall beautiful blonde people" who might have been male or female, dressed in uniforms, and who were "benevolent and empathetic." He was then introduced to the "Boss," who was more reptilian in appearance, and then escorted to a surgical theater where he was operated on by "Four to six grey alien-like beings." The subject was also "tagged" with an implant, and then returned to his home.[149]

The account is so similar to those encountered in the American literature as to suggest cultural contamination, but UFO and alien abduction reports are scarce coming out of India, and it has been suggested that this may be due to many such reports being classified as consistent with local folklore and religion and not associated with the UFO Phenomenon. For instance, sightings in a village called Ranichauri are common, and the beings there are referred to as *pariyaan*, or fairies.[150]

Another report, near Chandigarh, also discusses a four-foot-tall, gray alien with "big, black eyes." In another case, an Indian Ufologist claims he is in telepathic contact with aliens.[151]

Alien Phenotypes—Southeast Asia

Southeast Asian alien reports tend to be quite different. Here we find the "aliens" to be very small, in the range of three to six inches to only two feet in height. Their craft are similarly small, the size of large toys. Ahmad Jamaludin is a UFO investigator based in Malaysia, and he writes for the *Flying Saucer Review*, published in the United Kingdom, and for the *MUFON Journal*. He has been collecting UFO and close encounter reports from Malaysia, where there have been frequent sightings. In fact, one of us (Levenda) lived in Malaysia at the time of one such sighting and watched a video of the cigar-shaped object on national (state-controlled) television.

Several of Jamaludin's reports are of small-sized aliens. One sighting, near the town of Kuantan in the mid-1970s, was of a three-inch tall, brown-skinned or brown-suited being, which nonetheless looked human and seemed to be armed.

Jamaludin goes on to state:

> As far as humanoid encounters are concerned, there is something strange going on in Malaysia. All the entities encountered have measured 6 inches or lower. There is not a single reported encounter with the UFO occupant more than 6 inches tall.[152]

He continues by saying that 90 percent of the reported encounters have been by schoolchildren. This raises a number of issues. In the first place, can hallucinations or hoaxes be ruled out? In the second place, if we accept the reports as true, then we may have an important link between ideas of indigenous beliefs and reports of tiny beings—fairies,

elves—and the UFO Phenomenon. Of even greater interest would be connecting these Malaysian sightings with
reports concerning poltergeist phenomena in other parts
of the world, which seem to be linked to the presence of
pubescent and pre-pubescent children in the homes where
the "infestations" occur. Do children have a special capacity for sensing (or perhaps causing) paranormal events? The
Phenomenon has a definite consciousness component; are
we missing the boat by testing only adults for ESP and PK?

In another article, Jamaludin notes a variety of alien
encounters in Malaysia in the 1970s, and they all ranged
in height from three inches to six inches. All were seen
by schoolchildren with the exception of a sighting in
Sarawak (on Borneo) by several people, including adults
on vacation. They saw six or seven creatures, also tiny, who
appeared to be wearing white suits. Some had long hair,
and were assumed to be females. They were identical to
humans "except for their tiny size."[153]

In 1995, Reuters reported the sighting of a "huge UFO"
in Malaysia, in Selangor state (which is near the capital,
Kuala Lumpur). The occupants of this craft were described
as having "long ears and little red eyes." They were said to be
about sixty centimeters (two feet) tall, which would make
them considerably taller than their 1970s counterparts.

The Malaysian Air Force dismissed the sightings as
hallucinations.[154]

Abduction cases in Malaysia, however, seem not to
involve a UFO sighting. In these cases, the "entities involved
do not come out of UFOs. They seem to have stepped out

of a co-existing world, if indeed there is one."[155] There were no surgical operations, and no messages for the outside world were given. However, the aftereffects were similar to reported abduction cases in North America, with "temporary amnesia, extreme thirst, tiredness, and generally feeling emotionally upset." The entities dressed like local people but were obviously not locals, and the purpose for the abductions was unclear. It was, they report, like a kind of excursion to the other world.

Outside of Malaysia, sightings of aliens were usually quite different. In the Philippines, they are tall—about six feet—and appear Caucasian, but are also dressed in white suits like some of their Malaysian counterparts. They are associated with UFOs and in some cases are seen as exiting a "saucer"-type craft. Communication between the entities seems to be by gestures rather than sounds, further reinforcing the idea that the "aliens" (whether of North America, India, or Southeast Asia) do not converse the way we do, but use a different form of communication. The absence of sound-based communication may indicate a lack of ears and mouth as a baseline phenotype characteristic (regardless of the apparent height or skin color of the "aliens"), and thus helps to explain the large, outsized eyes of some exemplars as a primary means of sensing the exterior world.

Alien Phenotypes—Zimbabwe

Perhaps the most famous African event is the Ariel School sighting in Zimbabwe, which took place in 1994.

This is considered a seminal event in UFO history, as it involved a number of schoolchildren of different ethnicities. It has been researched and analyzed by many popular Ufologists—including famed abduction researcher John Mack—so we will not take a great deal of time with it here, but will instead focus on the physical appearance and attributes of the "aliens."

The sighting took place during the morning of September 16, 1994, at an elementary school in Ruwa, about twenty-two kilometers (fourteen miles) southeast of the capital city of Harare. Sixty children witnessed the event, and their ages ranged from six to twelve years old. They reported seeing several vehicles land, one large and a number of smaller ones, and two beings were seen outside the vehicles.

These "aliens" were described as black, or as dressed in black, and according to some witnesses one of them had very long hair. (This is an interesting deviation from most accounts from North America and other regions, which describe the "aliens" as hairless and their skin as leathery; very nearly amphibian in nature.) Their heads were described as long and their eyes quite large, with the usual spindly arms and legs. They communicated in some cases, but always telepathically.[156] They were short, perhaps a meter or a little more (from three to four feet) in height.

The students themselves were both boys and girls, of several races and ethnicities. Many were Caucasian (of South African or British ancestry, but Zimbabwe born) and others were "black African children from several tribes" as well as "Asian," i.e., children of Indian ancestry.[157] Video interviews taken a few days after the sighting show the racial and cultural diversity of the witnesses.

The Ariel sighting has been criticized, usually on the basis of the interviews themselves which—the critics suggest—made use of leading questions to encourage the children to fit their experience into a typical UFO narrative. That criticism is used to "debunk" the entire experience as a hoax or some type of collective hallucination.

However, a detailed contemporaneous published report casts doubt on the "hoax" theory. A local UFO researcher—Cynthia Hind—had been publishing a newsletter called *UFO Afrinews* at the time. Based in Harare, she was only a few miles from the Ruwa site and was able to talk to the students and teachers and to compile a report in two consecutive issues of her newsletter. It is clear from the report that the beings were observed by children of different ethnic and racial backgrounds, both boys and girls, and that the sighting generated feelings of both fear and excitement. In one case, it seemed as if one of the beings was staring directly at one of the children, who reacted with terror as if she was going to be taken or kidnapped in some way.

What seems consistent in the accounts is the physical description of the beings. In fact, even the discrepancies among the eyewitness accounts are such as to further reinforce the sincerity of the witnesses as well as the fact that they each saw something very unusual. For instance, most of the children described at least one of the beings as dressed in black, but one or two children claimed the being had black skin. This type of discrepancy would be normal if it were a crime scene account. The excited emotional state of the witnesses and the "high strangeness" of the event would contribute to slightly different accounts.

Of the telepathic communications, some of the children reported a series of horrific images of the pending destruction of Earth due to human causes, including war and pollution. These are recognized as the almost standard "warnings" received by contactees all over the world for decades. The purpose of these communications is unknown, of course, and we can only speculate. Are they actually warnings about specific future conditions, or is their purpose something altogether different, such as the creation of a state of high anxiety that would perhaps distract the witness from a deeper examination of the event as it is taking place? A kind of shock tactic, perhaps, or a measure designed to render the witness helpless in the face of so much horror? "Resistance," they may be saying, "is futile." Or are these images created by the human mind as a kind of mental reaction to what appears to be an existential threat posed by the very presence of the "alien"? A threat on many levels beyond the physical?

Again, as we have been insisting all along, it is useless to try to ascribe human motivations or human values to the actions of the Visitors. We have no way of knowing if these are indeed extraterrestrials or ultraterrestrials or some other type of image, being, or force. Our initial approach should be to record as accurately as possible, and then to collate the data received from both the UFO/UAP sightings and the close encounter experiences and to treat all of these reports seriously.

The Ariel School case provides an excellent opportunity to do this, because the event took place in 1994 and several of the witnesses have been re-interviewed more recently, as adults, to find out what they remember of that event, whether they have changed their story, etc. As mentioned, they were from an intersectional mix of genders, races, and ethnicities. The diversity of the eyewitness pool is particularly interesting because it enables us to look for consistencies and inconsistencies across the board.

The diminutive height of the aliens is consistent with most sightings around the world (with a few exceptions, as noted), as is the menacing or sinister aspect. The lack of verbal communication is another. Add to this the equally consistent reports that the aliens either emerged from, or were associated with, a machine or machines of some sort that to some of the witnesses appeared to be flying at treetop level before settling down in some brush. The beings engaged in bizarre behavior, wandering around or jumping, and seemed to make threatening gestures toward the children.

There were no adult witnesses to the craft or the creatures, but they were quite aware of the emotional reaction of the children who tried to tell them that something strange was going on outside the school. The principal of the school assured investigators that the students had seen *something*, but of course would not state that what they saw was a UFO or aliens, etc.

There was a "folkloric" connection to the event as well, with the "black African" children expressing fear that the creatures were going to eat them. This was reported as a reference to local traditions concerning the *tokolosh* or *tikoloshe*: a malevolent spirit that can cause death after being evoked by a magician. It is specifically considered a threat to schoolchildren. Thus, we are back in "chicken or egg" territory: did the beliefs concerning *tokolosh* influence some of the children who claimed they saw aliens, or are the *tokolosh* themselves an indigenous reference to aliens?

The Possibilities

Several "alien" characteristics stand out. In the first place, whatever these images ultimately represent, they don't represent human beings. That may seem like an obvious conclusion to draw, but it's deeper than that.

We, as humans, have no other experience when it comes to confrontations with the Other except in the form of phenotypes or body types that have a *few* elements in common with us but in the end not many. It would seem we need some point of reference, and that is usually a torso with a head and two arms and two legs. After that, the image of the "alien" or the Other can be anything: very small, somewhat small, relatively tall. Gray, black, blonde. Menacing, or benevolent.

We don't see images that are so foreign to our idea of conscious, sentient life that words fail us. We see something that looks vaguely human or humanoid or "human-ish," but we "know" that they are not human at all. It's not merely the high-technology aspect of the vehicles or machines that are foreign; it's also the passengers and crew.

Even then, though, the craft—saucers, steampunk airships, cigar-shaped vehicles—are comprehensible as transportation devices. They fly through the air. They deposit organisms that look a little like us, but not enough to fool us into thinking they *are* us. In fact, they are most definitely *not* us. We do not have three-inch-tall people living in our world; we don't have malevolent-looking, three-foot tall Grays with large eyes, either.

But why aren't they more different in appearance? Why not cephalopods, like an octopus or a squid? Insect-like?

No, no; that's still too Earthly.

Why not beings whose appearance would strain anyone's vocabulary to describe? Balls of glowing hair, for instance? Some elemental configuration halfway between water and steam? Something out of the imaginations of science fiction and fantasy writers? Something dimension-challenging and visually arresting, like a cross between H. R. Giger and M. C. Escher? We obviously can come up with images like these; we do it all the time in movies and television, in comic books and graphic novels. But the experiencers don't, even though they have had ample opportunity to do so considering the proliferation of "outer space" creations by our most talented professional "imagineers." So why are we still stuck with Grays and Nordics? And the odd Reptilian?

One of the criticisms leveled at those who claim they have seen aliens is the prevalence of Hollywood-style ideas of what aliens look like. It is claimed that those images inspire the actual reports, thereby rendering them ridiculous or worthless. If that is so, why hasn't anyone reported being abducted by either *Alien* or *Predator*? How come no one has reported seeing Spielberg's *E. T.*? Isn't it about time?

We are forced to conclude that the taxonomy of the actual "alien" is relatively consistent across times and cultures due to a similarity of genuine *experience* and not a flight of fancy or an act of the imagination.

That brings us to the next possibility.

If the Grays, for instance, are machines—androids of some kind, or robots, some "strong AI" device—then why do they look like humans, albeit rather warped-looking humans? If an alien civilization had created these devices for space exploration—and done so from a platform light-years distant—why would they have chosen a phenotype so close anatomically to Earth humans? If they already knew what we looked like and designed a machine to imitate that look (perhaps as a way of reassuring us) then why bother visiting Earth at all? And, more important, why not design the machines to look *exactly* like humans? Why the deviations? And why those *specific* deviations: no ears; no reproductive organs; no nose, or vestigial nose; no mouth, or vestigial mouth, etc.? The Grays do not breathe. They do not eat or drink or even speak. So why the weird camouflage? It's like wearing jungle fatigues in the snow.

We are left with a conclusion that really no one wants to hear.

▼   ▼   ▼

In the first place, if the aliens as perceived are actual, tangible beings and not holograms or projections of deep, unconscious human fears, then they were created by a race of beings that share a great deal of genetic material with us. The Grays may be machines, but machines of a type we have yet to invent: a cyborg that is two or three revisions beyond our current wildest dream of a cyborg. That is, it is part manufactured device and part organic being; however, the "manufactured device" element is an organic development of what was originally nonorganic material, deliberately designed with specific vulnerabilities to ensure (a) that it does not operate too independently or autonomously and (b) that it cannot reproduce itself.

If, however, the aliens as perceived *are* artifacts of consciousness—like congealed fears or coagulated dreams, projections into the visual field from areas of our minds that are cries for help from our collective unconscious— then we must confront the possibility that the human race is on the verge of a collective nervous breakdown, and the UFO sightings and the alien abductions are the canaries in the coal mines of our cerebral cortexes. You see, we— in the twenty-first century—are facing a situation that has not occurred previously in recorded history. We have the Internet. We communicate instantaneously with each other, facts and fictions, realities and fantasies, all at the same speed and with the same mindless conviction. We are in the process of creating a group mind, and the promoters of "strong AI" want their creations to plug directly into that group mind.

But what if our group mind is insane?

▼   ▼   ▼

Ufologists have criticized other Ufologists for promoting an idea that the "aliens" are demonic or evil or just plain dangerous to human beings. Human beings are extremely dangerous already. The events of the last century should tell you that. Yet that has not stopped us from making tremendous technological advancements.

Another race of beings from elsewhere could be just as technologically advanced as we are—or even much more advanced—and still be as dangerous as we are. We have no grounds for believing otherwise, as much as we would like to. At what point does technological advancement give rise to ethics, morals, and spirituality *in the human sense of those terms*? It hasn't so far, not for us. Why would we expect an alien race to have bested us in that regard?

Naturally, we hope that they have. If not for our sake, then for theirs. But can we take that chance?

Based on the anatomical design of the "aliens" we know that we are dealing with a phenomenon that taxes the best of our abilities to understand, even to recognize it for what it is. It's just human-looking enough to keep us guessing.

The fact that the "aliens" do not take on the appearance of talking trees or singing clouds or parked cars may reveal more about their true nature than they would like us to know. They may be restricted in ways that we can only guess at, *but restricted they are*. Grays, Nordics. The odd Reptilian. Playing with perverse human eugenics fantasies, perhaps. Encoded racism. Playing up fears of the Other to divide and conquer. Seduce with "beings of light"—like the ancient gods and goddesses—and then come in for the kill with angry, jealous gods of war and vengeance, of blood sacrifice and demonic possession.

"Why have they not invaded?" people ask. "Why have they not simply come in, destroyed our weapons systems, put their webbed foot down, and forced peace on Earth?"

This is why: because they can't.

We are in a symbiotic relationship with the alien. We have always been, since that first mention of the sons of God and the daughters of men. They are not all-powerful; they are just powerful in ways we are not. And they need us for that reason.

And, God help us, we need them.

HOUSE GUESTS

I N THE PREVIOUS CHAPTERS WE HAVE LOOKED AT A FEW
seminal ideas of the past hundred years where the
Phenomenon is concerned. It is obvious that there are
so many filaments to this particular web that to try to keep
our narrative purely linear is difficult and in the end inef-
fectual. Every time one strand crosses another, a whole new
area of experience is introduced and we are forced to follow
that strand back a little way in order to get some perspective on the whole.

This volume is entitled *Man*. That means both the arrival
of the Other into our world, *and our arrival into theirs*.
The twentieth century saw the arrival of human beings in
space as well as the arrival of what we call—without a lot

of confidence—"aliens" onto our planet. Space became a point of tangence between two realities. It became, for the first time in recorded history, a contested domain. The more we sent rockets and missiles and eventually animals and humans into space, the more "something" began to make itself known to us here.

As we said at the beginning of this project, there have been Visitors since time immemorial. The records of our ancestors all over the planet attest to that. Now, however, we have chased them, shot at them, photographed them, and tried to engage them on their own field of operation.

There are those who have been at pains to warn us against having anything at all to do with them.

But is that any way to treat a guest?

For the most part, the Phenomenon is a random series of appearances, visitations, even abductions. It happens to *us*. We don't happen to *it*. There is probably no other human experience that comes close to making us such passive observers/victims, except for some natural disasters. One thinks of an earthquake, or a volcanic eruption. Or a tornado.

Or rape.

In the case of alien abductions there is a definite sense of physical violation that commonly is reported. Human beings are removed from their environment and subjected to a variety of unpleasantness, mostly centered on their bodies but also extending to their psyches. This is a localized, microcosmic form of what takes place on the more macrocosmic level, with foreign spacecraft perceived as entering Earth's atmosphere at will—either from space or from some alternate reality—and "penetrating" our own.

The sightings of strange flying objects are often accompanied by a reaction that is more than simply noticing something out of place. Seeing a UFO exerts a strange fascination—even a sensory dislocation—over human consciousness. According to the literature, and based on hundreds if not thousands of eyewitness reports, seeing a UFO changes the witness's perception of reality altogether. Sometimes this change is relatively minor, a sense that the event exists in its own "moment" outside of the general flow of time and space. An anomaly. A wrinkle in the fabric of one's experience. At other times the change is more profound, a challenge to one's worldview, an inescapable sense that reality is fragile and that something serious has taken place that cannot be ignored or forgotten. A trauma. And indeed, it has been understood as such.

We are sleepwalkers here, on this planet. Our governments do not know what this phenomenon really is, so rather than admit ignorance, they deny its existence. This has been possible for decades, especially when the masses of people were unable to communicate directly with each other across the country and around the globe. Now, however, things are changing, and it will not be possible to keep our heads in the sand for much longer. We have ignored the Phenomenon for so long because it has not been perceived as a threat to "national" security, even as it has been a threat to personal security in so many cases.

J. Edgar Hoover famously refused to believe that the Mafia existed. He and the FBI could proceed fighting crime and enforcing the law without having to recognize the existence of organized criminal gangs with pedigrees going back generations in time and across the world. We are in a similar situation with regard to the Phenomenon. We can continue to ignore its existence as long as we address the localized forms of it on a case-by-case basis, throwing our limited resources behind a strategy of debunking individual cases or calling them "unknown" and moving on, and of ridiculing those who would take a different view. The problem with that approach—like Hoover's problem with the Mafia—is that the essential characteristic of the threat, the organizing principle, goes unrecognized and therefore unaddressed. The soldiers and button men (the assassins) are arrested, but the *capos* (the ones who give the orders) go free. Thus it is with the Phenomenon.

Critics of the alien abduction experience, for instance, claim that amateur or nonprofessional therapists and hypnotists have created the concept through the implantation of false memories. This has been challenged many times, by many experts. Transcripts of therapy sessions made available in a number of books and reports indicate otherwise. What we may be dealing with are not false memories but repressed memories. In that case what we are experiencing is *an entire planetary population that has repressed its collective memory*. The constant barrage of UFO sightings around the world in the past seventy years, accompanied by insistent reports of individuals having experienced contact with alien beings, including abductions, physical and sexual molestations, and ancillary events, may be symptomatic of a widespread, near universal mental disorder that is the result of a primary trauma. Our brains, our psyches, are starting to wake up from a deep, troubled sleep. Elements of our nervous systems are coming alive in unforeseen ways, reporting experiences for which we have no language, no vocabulary, save that of the alien, the extraterrestrial, the nonterrestrial, the Other. We are beginning to sense the Mafia in our world, the "men in black" who haunt the planet like the ghosts of a dead civilization that once lorded it over our mountains, our deserts, and our plains. Except it is not they who are dead, nor are they the ghosts.

It is us.

The Phenomenon is awakening us to new perceptions and creating cortical fissions and fusions in neural pathways, reengineering our consciousness, rewiring our brains. We may be slowly awakening from an unconscious state, rising from our graves into the waking world. Even Carl G. Jung, when addressing the UFO Phenomenon, likened it to a process of collective individuation with the image of the "flying saucer" a mandala for a modern world.

But is it a mandala, or a magic carpet?

What of those among us who actively seek contact? Is it possible to reach out, to invite the Visitors into our plane of reality or even to gate-crash their world?

The practice of channeling has provided us with a metaphor for making contact with the Phenomenon. Formerly known as mediumship, channeling is the use of a voluntary human intermediary between our world and some Other. While mediumship was most famous as a method of contacting spirits of the dead—or in the case of Dee and Kelley described in a previous chapter, of angels—channeling extended the reach of the medium into other realms. Andrija Puharich held séances in 1952 to channel the extra-planetary forces called "The Nine": these were not spirits of the dead, but beings who identified themselves as gods and, eventually, as beings aboard an invisible spacecraft in low Earth orbit.

It was Whitley Strieber who began to write of seeing dead people as an adjunct to the UFO and abduction phenomena. Does this congregation of paranormal experiences—mediumship, channeling, UFOs, ghosts—reflect something important, a window into how our reality is structured, or is it simply the muddled thought processes of those who have been touched by the Phenomenon in a way so profound as to inspire the urge to see *everything* through the lens of contact? How do we "discern," as the Evangelicals say, one type of spiritual or paranormal entity from the other? Angels, demons, gods, jinn, Small Grays, Tall Nordics, Reptilians, insectoids, humanoids . . . aren't they all just machines, devices we've created as drones (as unmanned *psychological* vehicles) to penetrate into hidden realms of our own consciousness?

Well, they may be drones, these sekret machines, but are they *our* drones?

We in the West, we Westerners, are obsessed by the difference between fiction and nonfiction. We think we can tell one from the other. The Phenomenon challenges that very basic (to us) assumption because there are actual *machines*. They appear on radar screens. They are photographed. They are seen by people who are awake, in broad daylight. They disrupt communications, electrical transmissions, and fuck around with missile bases. But there is an element of these machines that screams *fiction.*

Some languages do not have words for "fiction" and "nonfiction." There are countries where biographies of famous people are shelved in the fiction section. Whatever is a narrative is considered fiction in these places—but not fiction the way we understand it. Not as lies, or deception. Works of the imagination, maybe, which is not the same as a lie or deceit.

We've struggled with this distinction, even in the West. Truman Capote and Norman Mailer wrote "nonfiction novels" in an attempt to erase the difference or, at least, to create a new genre—one that would partake of the nature of both fiction and nonfiction, of narrative and textbook, of poetry and engineering manual. They knew that something fundamental had taken place with the scientific revolution and the resulting bifurcation of the human mind: the creation of an authority that would cast the deciding vote between what was real and what was not, what was nonfiction and what was made-up fiction.

What was science, and what was religion.

Three hundred years ago a no-man's-land was created between the halves of the bicameral brain, a line drawn along the corpus callosum. A wall went up. Barbed wire neurons. Watchtowers. Armed guards. Dead refugees.

"Mr. Newton, tear down this wall!"

The Phenomenon exists *there*. In that spot. The place where the house guest comes to stay for a few days. The guest room: a place that is uninhabited by the family who lives there. A place set aside for transient phenomena. The extra room in the memory palace, the one no one uses. It has its own entrance, in the back. A hot plate. Clean sheets. Dust.

And then, suddenly, an arrival. A relative. One rarely seen. *But related by blood.* With maybe a gift from abroad. Something exotic. A disruption in the daily pattern of behavior. Dislocation. Something . . . something slightly . . . disturbing, maybe. Different.

And then, just as quickly, it's gone; the relative has departed for other lands, parts unknown. And the house seems empty, and quiet.

But the gift has been left behind. It sits there. In the dark. In the guest room.

▼   ▼   ▼

Ticking.

Galileo's Revenge:
Science versus Spirituality

IN THIS VOLUME WE DESCRIBED THE CURRENT STATE OF investigation and revelation concerning the physical and scientific aspects of the Phenomenon. This required us to address some of the more serious questions aimed at those who study and research UFOs and related aspects of the Phenomenon. This especially required us to confront the objections to any serious study of the Phenomenon raised by atheists and some scientists, as they seem to control much of the mainstream narrative.

It is a fact of academic life in the United States (and in many other countries) that governments and the military are the only human institutions with the necessary economic resources to finance scientific research and development.

Even corporations are funded by government grants and contracts that enable them to survive. This results in a relationship between governments, the military, and corporations that President Eisenhower famously characterized as the "military-industrial complex." In fact, it is entirely possible that the power of governments is circumscribed by the military-industrial complex to the extent that control of information concerning the Phenomenon is carefully monitored. As we saw earlier, much of the serious research into the paranormal—for instance—was conducted in the United States under the auspices of the military and the intelligence agencies, which had a serious interest in the weaponization of the paranormal. This Cold War approach is noticeable in the engagement of the military, the CIA, and the FBI with the UFO Phenomenon as well. In fact, they were often seen as interrelated, which is why we have grouped all of these concepts under the general rubric of the Phenomenon.

This and the first volume of this series have laid the groundwork for what follows in the third volume, *Sekret Machines: War*. In that book we will address evidence for the Phenomenon and the reception and interpretation of that evidence by a variety of world civilizations and will provide an in-depth examination of the explosion of this research, which began at the very end of World War II, and the effect the Phenomenon had on the origin and development of many seemingly unrelated events in politics and science. Based on this information we will proceed to an analysis of what we have learned and will offer recommendations and suggestions for how an enlightened society should approach what is most probably the greatest single issue facing twenty-first century civilization.

It is hoped that this project will inspire a new generation of scientists, engineers, researchers, philosophers, psychologists, astronomers, geneticists, and artists—people from the hard sciences, the soft sciences, and the humanities—to reevaluate the evidence and to reimagine our society as a space that includes some of the seemingly fantastic (and "irrational") elements of human experience integrated with its more rational and familiar aspects in order to create new ways of understanding the complex matrix of knowledge and experience, of intellect and emotion, of artistic and scientific creativity, that we optimistically call "reality." Before it does, however, we have to address the ideological conflict that exists between scientists and experiencers.

> Monotheism does not really like the rational work
> of scientists.
>
> —Michel Onfray[158]

> Religion is the sigh of the oppressed creature, the heart of a heartless world, and the soul of soulless conditions. It is the opium of the people.
>
> The abolition of religion as the illusory happiness of the people is the demand for their real happiness. To call on them to give up their illusions about their condition is to call on them to give up a condition that requires illusions.
>
> —Karl Marx[159]

One would think that the alien astronaut theory would be a congenial one for atheists. By viewing *all* religion as the survival into modern times of an ancient cargo cult inspired by alien contact in the mists of history, the fact of religion and religious "delusion" would be explained in a way that would make atheists more comfortable with the idea. At the same time, it would satisfy a growing number of people—especially non-scientists, even anti-scientists—that there is "something" to religion and the religious experience that can be addressed using the rubric of ancient aliens and ancient astronauts, which retains a vaguely mystical and otherworldly concept. By replacing "God" with "Alien" we would remove the supernatural from the equation and replace it with something that is at least scientifically possible: the arrival at some point long ago of an alien being that triggered everything that characterizes popular religion. We could explain all the attributes of a God in the "heavens"—superhuman power and intelligence, super weapons, ascending on clouds, miracles, etc.—using only the idea that there was alien contact with human beings, and that the attempts of those humans to describe the event using a limited vocabulary that predated modern scientific explanations is what resulted in what we know as religion today. It seems that this would be a perfect solution for atheists and a way to soften their stance against the believers by saying, in effect, "Yes, there was a seemingly supernatural occurrence long ago that was interpreted as divine intervention. It is easy to see how people could have become confused in those days, but the scientific discoveries of the past fifty years or so are leading us toward a different explanation for these attempts to describe the ineffable in spiritual terms. You are not deluded at all; your belief system is based on actual events that took place long ago and that we are just now coming to understand."

Alas, such a reasoned stance was not to be. At least, not yet.

Opponents of religion are opposed to anything that smacks of the paranormal. They seem fearful that acknowledging the paranormal opens the door to religion, letting a disembodied deity into the back room. While atheists blame religion for violence and division, with some justification, it is hard to identify a time when astrologers or alchemists, mediums or psychics, waged holy wars (but let that slide for now).

The problem of science versus religion is in sharp relief when we confront the subject of UFOs and the Phenomenon. When we listen to the narratives of people who have seen UFOs we realize we are not dealing with people who are willful participants in a delusion, or promoting some form of religious dogma or mysticism. These are ordinary people from all walks of life—including trained observers such as military officials, pilots, and radar operators—who are suddenly and without warning confronted by something they cannot explain. Moreover, that "something" exerts an effect on their consciousness such that the experience is combined with feelings of awe, fear, paranoia, confusion, etc. No one who has been an "experiencer" can ever forget the moment when they confronted

the Other for the first time. No one who is not an experiencer can understand the profound emotional state that
ensues. Yet, those people who have encountered the Other
are informed—in what passes for scientific piety—that they
did *not* see what they thought they saw; they did *not* experience what they thought they experienced. They are told
that the reason for their experience ranges from a mistaken
sighting of the planet Venus to swamp gas, radar anomalies, night terrors . . . all the way to outright deception or
hallucination. In other words, the experiencers are heretics.

Science, which usually proceeds from data and observation and works backward from there to develop working hypotheses, begins instead with the premise that there
is no such thing as whatever it is the experiencers claim.
Therefore, the experience is invalid and the experiencer an
uneducated, ill-informed rube, at which point the skeptics
man the barricades and throw stones at both the experience
and the experiencer in an *intifada* of derision.

If this seems unduly harsh, one only has to review the
available literature.

It is a truism that books written by skeptics attacking
UFO sightings and the paranormal generally do not sell
very well. No one is titillated by debunkers, and one can be
forgiven for feeling that those who do enjoy reading books
by skeptics are members of a small group of self-congrat-
ulatory cynics who pride themselves on representing a
secular version of "holier than thou." Certainly that is an
unfair characterization of a very serious problem deserving of patient and sober analysis, especially when we find

ourselves forced to confront a phenomenon that is resistant to calm rationality.

However, there is a kind of scientific inquisition at play when it comes to the paranormal, with modern equivalents of Torquemada and Savonarola who go out of their way to prove the weaknesses and probe the vulnerabilities in the claims of paranormal researchers, psychics, UFO experiencers, mediums, astrologers, "ancient alien theorists," and others, declaring them adherents of a kind of heresy: a deviation from "truth." It is almost an obsession, and one wonders who they imagine their readership may be. The debunkers seem to assign to themselves a cultural (not only an intellectual) superiority, and it is only one step removed from debunking the paranormal to attacking religion in general. The debunkers of the paranormal—in their allegiance to what they believe to be a scientifically based perspective—are atheists, or at least fellow travelers. They are the cultural colleagues of Richard Dawkins, Christopher Hitchins, Michel Onfray, and Sam Harris, and the heirs of a tradition that goes back at least to Galileo.

▼　　▼　　▼

In 1616, the Roman Inquisition found Galileo Galilei (1564–1642) guilty of heresy due to his insistence that the earth revolved around the sun, thus putting the sun at the center of the planetary system. The Inquisition had decided that the heliocentric view supported by Galileo and based on the Copernican system was scientifically and philosophically untenable as well as heretical, since it contradicted the Biblical account of Creation as it appears in Genesis as well as other Biblical references to the sun rising and setting, etc. Galileo was sentenced to house arrest for the remainder of his life and was not allowed to publish. He continued to write, but his writings did not see the light of day until long after his death. Incidentally, his trial occurred sixteen years after the Inquisition ordered the philosopher Giordano Bruno burned at the stake for holding similar ideas.

This episode is famous as the last gasp of an anti-scientific religious establishment—the Roman Catholic Church—in the face of scientific discovery. Since the Copernican-Galilean system is now accepted as true, it can be said that the victory of the Church over Galileo was the victory of lies or ignorance over truth and knowledge.

As the scientific revolution took hold in the decades and centuries following Galileo's death, the scientific community understood that they had to deal with an antagonistic faith-based environment that saw their research and their discoveries in danger of being characterized as heresy or witchcraft. However, the shift in emphasis from the earth to the sun represented by Galileo had far more important repercussions than merely a critique of the Biblical references: it shifted the focus of creation itself from the earth (with its human beings made in God's "image and likeness") to the vague center of a vast and dark universe. It made of humanity little more than an artifact, relegating humans to a spot in the cosmos that is the third rock from an insignificant star in an insignificant corner of the Milky Way galaxy, which is, itself, only a minute fraction of the universe's 170 *billion* galaxies. The implication was profound, and profoundly unsettling.

It is this relegation of human beings to the basement of the cosmic mansion that has contributed to the kind of existential angst that began to develop in the eighteenth and nineteenth centuries and especially with the philosophy of nihilism and, of course, existentialism itself. Hegel, Schopenhauer, Darwin, Nietzsche, Kafka—all the way to Freud—contributed to a view of the human race that has no real purpose, no real meaning in the world, other than survival: the genetic imperative. It was Nietzsche who famously claimed, with no little horror mixed with manic glee, that "God is dead." It was, we suggest, men like Galileo and his successors who were the perps, gunning God down in a Florentine drive-by; it was Nietzsche who found the body.

The church knew this was happening; they saw the Biblical handwriting on the wall. Galileo's telescope had enabled priests and cardinals, no less than astronomers and physicists, to gaze upon the face of the Moon or to calculate the phases of Venus or to count the moons of Jupiter. There was a new religion abroad in the land, a heretical faith that would challenge Holy Writ from a completely unique perspective: logic and rationality, devoid of dogma and the received or revealed wisdom of a complex, almost gymnastically articulated theology. It should be remembered that the church represented the Establishment in

seventeenth-century Rome as Galileo languished in isolation; soon, the growing scientific establishment would challenge not only the church but all religions and even the idea
of religion itself. How could one be a person of faith *and*
a person of science? How could one believe in the myths
and fairy tales of the world's scriptures while simultaneously
calculating the orbits of comets or the number of microbes
dancing on the head of a pipette? To hold both points of
view—the religious and the scientific—in the same brain,
in the same mind, is considered by professional atheists such
as Sam Harris to be tantamount to schizophrenia, madness,
sheer insanity.[160] One has to choose, and in the absence of
miracles and divine interventions (both in short supply in
the twenty-first century *anno Domini*) it seems safer to side
with the science that gave us microsurgery, microwave ovens,
and weapons of mass destruction. You can't beat them, so
you might as well join them. And maybe get a flat-screen
TV in the process. All you have to do is admit that religion
is stupid, murderous, and the playground of the dishonest,
the deceitful, and the destructive. Science gives you clean
water, fast cars, and all the technological advantages of high-
speed data, personal computers, Netflix, and smartphones.
Actually there is really no need to make a conscious choice.
It's made for you. And the distance between the scientific
mindset and the spiritual perspective grows larger with each
increase in processing speed, in data transfer rate, stretch-
ing until—it is hoped—it breaks completely and religion
is unceremoniously (ha!) tossed into the dustbin of history.

Galileo's Revenge.

Science versus Religion in the Twenty-First Century

The rise of religious fundamentalism—which is at least partly if not entirely a reaction to materialism, scientism, and technological advances and the superiority of those countries that boast of vast technological and economic power, not only in fantastic weaponry but also in popular devices and the ideologies that celebrate the cultures that gave them birth—is taken as evidence of the danger of religious sentiment and its associated irrational (and thus mentally disordered) consciousness. This conflict between science and religion is not only a struggle between reason and unreason, between sanity and insanity, but also between two completely different cultural attitudes. The smug superiority of the atheist and the scientist confronts the smug superiority of the religious fanatic. Both sides know they are right: the fanatic, because God is on his side; the scientist, because . . . well . . . in absence of evidence to the contrary.

The Atheist Manifesto, penned by Sam Harris,[161] is a case in point. This slender, nine-page document sets forth the position of the atheist vis-à-vis religion, and it is revealing in ways that its author perhaps never intended.

It is full of anger and vitriol, a less artful declaration than Nietzsche's, perhaps, but nonetheless insistent that all one has to do is look at religious fanaticism to understand what is wrong with the world. One would believe, reading Harris, that all violence comes from religion; that somehow a scientific attitude ultimately would lead to world peace. One can understand, reading Harris, that science—after so long reveling in Galileo's Revenge—now had to confront the idea that maybe victory was declared too soon. "Mission Accomplished": Not.

The roots of this problem are ignored. It is implied that modern fanaticism and its violent excesses have their origin in religion. This is demonstrably false. There have been fanatical atheists, such as Stalin, Mao, and Lenin; their collective atheist ecstasies resulted in the slaughter of millions of human beings, mostly their own. Harris rejects this argument, stating that the violence and brutality of, for instance, the Holocaust is an inheritance from German antisemitism that predates by centuries the rise of the Nazi Party. If his reasoning is right, then there is no hope for any of us, since we are all inheritors of racism, antisemitism, colonialism, and so on, and any evils perpetrated by any European, Asian, African, or American nation can be linked to its respective religious inheritance regardless of how secular or irreligious their government happens to be in the present. His rejection of this point about fanatical atheists such as the communist leaders mentioned above is that, although they were "irreligious to varying degrees, they were not especially rational." This seems to be closer to the point Harris is trying to make; that it is not atheism that is the model for a sane society, but rationality. For Harris, however, rationality and religion are polar opposites.

In trying to define what atheism really is, he complains that "no one ever needs to identify himself as a non-astrologer or a non-alchemist. Consequently, we do not have words for people who deny the validity of these pseudo-disciplines." There is a problem with this statement, of course. No one ever needs to identity himself or herself as a non-astronomer or a non-chemist, either. Further, a case could be made that astrology and alchemy are not "pseudo-disciplines" but "pseudo-sciences" inasmuch as anyone claims they are sciences. That they are disciplines there can be no doubt. Harris also refers to Judaism as a "heresy" in the eyes of the Catholic Church; this is a misuse of the term. The Jews are not Christian heretics; one could say that Christians are Jewish heretics, however. He also seems to misuse the idea of belief, insisting that one should believe things for which there is evidence. That would imply that one "believes" in evolution, for instance, or in gravity. These are not subjects for belief precisely because the scientific evidence exists to support them. One "believes" in things for which there is no evidence.

It is this cavalier tossing around of religious terminology that weakens Harris's argument, because it implies an impatience with the subject and an eagerness to have his say. It is a passionate attack on religion, a scapegoating of spirituality, avoiding the issue that the type of violence we are witnessing today would not be possible without the technological advances in weaponry that science has afforded us, from small arms to poison gas to dirty bombs. Moreover, it absolves—by omission—the rational, scientific, materialist communities of any culpability in the world's rampant terrorism and horror.

In the end the brunt of Harris's *Manifesto* is directed at the violence perpetrated in the name of religion, and in the complacency with which the pious greet everything from natural disasters such as Hurricane Katrina to the brutality of the Holocaust. Religion is identified as the culprit from the Israel-Palestine issue to the Catholic-Protestant conflict in Northern Ireland. Anyone following the news of the last fifty years could be forgiven for believing this. However, atheists take advantage of the lack of historical context among their average demographic so that it becomes easy to frame the rise of Al-Qaeda or the Islamic State purely in terms of religious fanaticism when in fact the roots of Middle Eastern turmoil are far more complex and have much more to do with European and American influence and control in the region—and, of course, oil—than with a narrow interpretation of the Qur'an. The Israel-Palestine situation is much more complicated than a religious conflict, although manipulators on both sides try to portray it as such. The roots of that conflict go back to the First World War and European adventurism and colonialism in the region. The violence in Northern Ireland was a consequence of English hegemony over Ireland; the English invaders were largely Protestant, and the Irish were largely Catholic. The English manipulated religious sentiments in order to maintain their dominance in the region, their last stronghold on the Emerald Isle. It was hardly a case of passionately religious Englishmen striving to bring their Protestant faith into Belfast, just as the British presence in India was not the result of Protestant missionaries desiring to convert the Brahmins, and their presence in Malaya not the desire to convert Muslims.

The attacks by atheists on religion are directed at what they see as the *implementation* of religion—violence, fanaticism, the subjugation of women—and not its essence. Attacks by the religious on science could easily point to its implementation as well: pollution, nuclear weapons, climate change, and the depletion of natural resources.

Of course, when the Japanese bombed Pearl Harbor in 1941, they did not do so out of religious motivations. And when we dropped atomic bombs on Hiroshima and Nagasaki in 1945, religion was not the reason. When Hitler invaded Eastern Europe, the Low Countries, France, and Russia, it was not done for religion but for hegemony and imperialism. The Roman conquest of Europe; Alexander the Great all the way to India; the Dutch in Indonesia; the Japanese in Manchuria, China, Korea, the Philippines, Malaya, Indonesia; the slaughter of the Native American populations of North and South America . . . none of these were done for religious reasons (although religion can often be cited as giving moral justification to those unsure of the morality of genocide, war, and enslavement) but in the name of economic and political expansion. Religious wars certainly existed, but the ones who benefited most from them were secular rulers: emperors, kings, and queens, who saw their territories expanded and their subservient populations increased. To focus on religion as the source of the world's unrest is to ignore these facts and to divert attention away from the role that secular governments and their scientific assets play in fomenting violence.

That atheists have a political agenda that may cause them to cut a few intellectual corners themselves should come as no surprise. While Harris, Dawkins, and Hitchens have thrown down the gauntlet on religion—especially organized religion—there has been for decades a sideshow conflict between science and the "irrational" that has gone largely unnoticed except for the participants themselves. Ironically, it may be in this particular sideshow that the greater conflict will find its resolution, particularly because the "other side"—the "irrational" side—has its share of scientists as well.

Science versus the Paranormal in the Twenty-First Century

There is a cottage industry in debunker literature, and since most devotees of the paranormal steer clear of books by skeptics, its claims are rarely challenged. These claims address the irrational in parapsychology, astrology, life after death, etc., and especially UFOs. There's nothing wrong with that. Even though debunkers and supporters of these ideas often talk at cross-purposes, there is merit in identifying logical inconsistencies wherever they may be found. That doesn't imply that the irrational must be made rational—"the crooked made straight," to cite a Biblical precedent[162]—because that indicates that logic and reason are the only metrics by which human experience should be judged, and there is no obvious justification for that point of view. Yet, in identifying logical inconsistencies we come

closest to an understanding of how the irrational is structured and why these ideas—from religion and organized
spirituality to astrology, alchemy, parapsychology, and the
UFO Phenomenon—are so powerful, even in the present
age of a digitally connected world. It is this very web of
scientific anomaly and physical "impossibility" that forms
the warp and woof of consciousness itself.

While atheists certainly have a point when it comes
to the hold that religion has over the human race and
the hideous results of much of that influence, the human
experience includes events that have no scientific basis or
support but that nevertheless occur. Ironically, this is precisely where both science *and* religion are at their weakest:
science, because the phenomena exist beyond any doubt;
religion, because the phenomena occur outside of any kind
of denominational framework. It is as if reality is saying,
"A plague on both your houses."

Atheists, when they address these phenomena, automatically associate them with religion and delusion. In
Carl Sagan's book *The Demon-Haunted World: Science as a
Candle in the Dark*, a brief chapter devoted to "Aliens" dis-
misses the Phenomenon in a very cavalier way. It is not an
example of careful debunking but a general polemic against
the idea that Earth has been visited by extraterrestrials based
on the fact that those who have observed the Phenomenon
do not have the same educational background as Sagan
himself; thus they must be mistaken. Sagan in that chapter
invites the reader to join him in pitying the poor fools who
believe in flying saucers, Roswell aliens, and the like, even

including those who have seen radar evidence (dismissing the radar traces as "anomalous propagation" caused by temperature inversions).[163] Sagan's approach in this chapter is general, not specific, except when it comes to debunking crop circles by focusing on two well-known British hoaxers and extrapolating from their example to an overall dismissal of the entire subject. The gymnastically articulated theology of the Catholic Church becomes, in the hands of Sagan and his colleagues, an equally gymnastically articulated critique of the Phenomenon. The only possible takeaway from these polemical attacks is an attempt to circle the wagons and devalue in advance any serious examination of these extremely important, albeit "anomalous" experiences and observations.

When conspiracy theorists accuse the US government of a cover-up where UFOs are concerned, they sometimes miss the point. A deliberate cover-up is not necessary when everyone with any knowledge of the matter gets their paycheck from the same place. Government funding of science and of research and development projects is such that no overt commands or policy statements are required. Everyone knows what they should do when they are brought up within a corporate culture that instills its own belief system in its members. It's as much unconscious as it is conscious: a knee-jerk reaction that might have been calculated and deliberate at some point in the past—in this case, based on a carefully considered policy crafted during the early days of the Cold War—but now has become part of the culture.

This has important implications for the science behind the Phenomenon, for if the military, or intelligence agencies, or the government itself has knowledge that is being withheld, either for reasons of national security or because the relevant agencies simply don't know what to do with it, then this knowledge is being withheld from civilian scientific circles as well. The critique that is usually leveled at UFO "believers"—that the Phenomenon as described simply could not exist because it violates known laws of physics—ignores the possibility that the Phenomenon does exist and behaves in accordance with certain permutations of physical laws in ways that are not yet understood because the evidence has not been made available. This is similar to the objections that were raised by scientists in the nineteenth century against the possibility of manned flight, for instance, or the transmission of radio signals without wires, etc.

To give two examples:

Tom had an interesting meeting in September 2015 with a very high-ranking military officer, someone with knowledge of black budgets and the inner workings of the Pentagon. In the course of this meeting, which was conducted in some secrecy at an airport in the Southwest and which danced around a number of hypotheticals, the officer conceded that if there was evidence of military knowledge of UFOs, any broad-scale effort to declassify files on this phenomenon was doomed to failure for purely bureaucratic reasons.

For instance, if one studies the research done in PSI and extrasensory perception, including remote viewing, by agencies of the US government over the past fifty years or so, it soon becomes very clear that this particular project went under various names and other identifiers (as we have seen) and was transferred from one department to another and then from one branch of the military to another, with various officers in charge at various times. This means that the study of Psi was not done by a single group or single government agency but by a number of disconnected organizations, for different purposes, and with a change of direction and emphasis each time. Different civilians and civilian groups were also involved, and that too changed from year to year. Add to that the usual environment of office politics and jockeying for position and you have all the elements necessary to obfuscate completely any substantive understanding.

When it comes to the Phenomenon, the situation is even more complex and diffuse. The modern UFO era began the same year as the Cold War: 1947. On June 5 of that year, the Marshall Plan for the rebuilding of Europe was announced. The House Un-American Activities Committee began to investigate allegations of communism in Hollywood. On July 26 of that year, President Harry S. Truman signed the National Security Act, which created the Central Intelligence Agency as well as the Department of Defense and the National Security Council.

And on June 24 of that year, seasoned pilot Kenneth Arnold reported the sighting of a squadron of unidentified flying objects near Mount Rainier in Washington State and inadvertently inaugurated the "flying saucer" meme. Thus, the advent of the UFO era occurred about a month before the CIA was established, and at a time of heightened tensions between the Soviet Union and the West.

In the months and years immediately following the Kenneth Arnold sighting, which was widely reported in the press, the Federal Bureau of Investigation began its own investigation of the Phenomenon. Agents on the West Coast and particularly in the Pacific Northwest were ordered to interview witnesses and collect information about sightings of unidentified flying objects. It was believed that those objects might be secret machines created by the Soviet Union to penetrate US airspace. There were also rumors that those machines might be the result of Nazi rocket technology, which had outpaced Allied attempts to create jet fighters during World War II. (At the same time, in Sweden, there were numerous reports of "ghost rockets" that had been buzzing Swedish citizens in remote areas of the countryside since 1946, something that was considered a possible attempt by the Soviets to intimidate the West by using captured Nazi technology. To this day, the true nature of this event has yet to be understood.)

So it is important to realize that the first publicized UFO sightings of modern times were taking place at the precise moment that the United States and the Soviet Union were on the verge of an armed conflict in Europe, a conflict that never became a hot war but was sustained as a cold war from 1947 almost to the fall of the Soviet Union on December 26, 1991. That meant that UFO sightings originally were being investigated as possible national security threats, and as such the investigations involved the newly formed Defense Department, the FBI, and local law enforcement agencies as well as the Department of the Air Force in particular (also newly formed as an entity separate from the US Army). Once it was determined that the observed aircraft were not an immediate threat, the focus of the military shifted to more immediate concerns, such as the Soviet Union's development of their own atomic and hydrogen bombs, ballistic missiles, and other war materiel.

▼　　▼　　▼

Advocates of Disclosure assume that a command from the president of the United States is all that is needed to declassify the UFO files once and for all. That is far from the truth. There have been so many different investigations over the course of more than sixty years, conducted by a wide variety of government and military agencies, that no complete file on the Phenomenon exists. There are many hundreds or even thousands of files containing information concerning UFOs spread and scattered through all of these sources, classified according to project names or numbers that differ with each organization and with each committee within each organization. Individual projects—such as the infamous Project Blue Book—can be declassified once identified, but a blanket declassification order is unlikely to produce any meaningful results. This makes getting American scientists involved in UFO research a particularly difficult task as there is no reliable information stream coming out of the military or the government containing useful raw data for analysis. Scientists who might be tempted to take a closer look are hampered by the fact that the research that does get published is often by special-interest groups who are promoting a UFO agenda, which leads to a suspicion that their data may be untrustworthy.

As Tom's source indicated, specific files relating to Cold War operations may contain UFO-related information but may not be accessible. For instance, there is the case of UFO-incited nuclear confrontations with the Soviet Union.

These were incidents that are well-known and in some cases amply documented by both sides. It would be foolish to dismiss them on the basis that our current level of scientific understanding denies their possibility. After all, we almost went to war over them.

▼　　▼　　▼

On October 5, 1994, and again the following year, the ABC-TV News program *Primetime*—hosted by Sam Donaldson—broadcast a report on a Russian UFO sighting. The report was filed by David Ensor, a seasoned veteran journalist who was based in Moscow for ABC News and who was involved in an investigation of declassified KGB files that were made available—in cautious releases over a period of time—after the fall of the Soviet Union in

1991. These files contained information on UFO sightings from all over the former Soviet Union, many of which were dubious, but a few stood out as being exceptionally credible. One of these was by a former lieutenant colonel of the Red Army, Vladimir Plantonev.

The incident occurred on October 4, 1982, in what is now Ukraine. The Soviet Intermediate Range Ballistic Missile (IRBM) base was located near the small town of Byelokoroviche, about 150 kilometers (94 miles) southwest of Kiev. The significance of this incident is due to the fact that two anomalous events took place simultaneously: one, the appearance of a large UFO over the military base; two, the "spontaneous" illumination of the control panels for the IRBMs that indicated that the launch codes for the R-12 nuclear missiles (the same type of missile that was being deployed in Cuba during the missile crisis of 1962) were being input without human involvement.

In October 1982, Ronald Reagan was president of the United States. He would famously refer to the Soviet Union as the "evil empire" in a speech in March 1983, and later that same month would announce the formation of the "Star Wars" Strategic Defense Initiative (SDI). This was a time of heightened tensions around the world, with an American administration that was determined to put an end to the Soviet Union once and for all. Thus the very strange initialization of the launch sequence for a sophisticated Soviet weapons system could have led to a confrontation with the West that would signal World War III and "mutual assured destruction."

Prior to the panels lighting up, however, a UFO sighting was reported by various Soviet military personnel, from radio operators to captains to a lieutenant colonel. The UFO was described as huge, about 900 feet in diameter and as tall as a five-story building, and was visible for hours. Virtually everyone at the base and in the town saw it, except for a captain and his team, who were down below with the instrument panels that were inexplicably lit up by an unseen force that began entering the launch codes. After fifteen seconds of absolute terror on the part of the missile operators the launch codes were reset back to normal: again, without human intervention. Thus the missiles were not launched and the world avoided a nuclear holocaust.

An abbreviated form of this report was aired by ABC News, as mentioned. The reporter, David Ensor, went on to become the director of Voice of America after a stint as director of Communications and Public Diplomacy at the US Embassy in Kabul, Afghanistan. Thus, the source of this report was not an impressionable young intern trying to make a name but a well-respected and wholly rational journalist who was trusted by the US government to occupy sensitive positions in foreign relations. That does not mean that his report on the Byelokoroviche incident is to be accepted at face value simply because he reported it, but it does indicate that his access to the KGB files and to the personnel who later went on the record concerning the event is most likely above reproach. This incident should be enough, in and of itself, to demand at least grudging acceptance by the scientific community of anomalous phenomena that violate known laws of physics and of the potential danger those phenomena pose to military installations worldwide. If this were an isolated case, it would still be important enough to require an in-depth scientific examination; however, it is not an isolated case. There were others in the Soviet Union and in the United States as well.

Perhaps the most credible of the American events is an incident that took place in the late 1960s at Minot Air Force Base in North Dakota. The event was described by David H. Schuur, a first lieutenant at the time and member of the Minuteman missile crew in the 455th/91st Strategic Missile Wing, in a series of communications with UFO researcher Robert Hastings.[164]

Minot AFB was the scene of UFO activity in 1966 according to the *Minot Daily News*[165] and based on a *Saturday Evening Post* article by Dr. J. Allen Hynek. This may have been related to the event described by Schuur in which he stated that the "Launch In Progress" indicators were set as the reported UFO flew overhead. Schuur had to hit the "Inhibit Launch" switches in order to keep the missiles under his command from being launched. Once the UFO had passed, however, all the indicators returned to normal.

This event is virtually identical to the Soviet version previously mentioned, except that it occurred fifteen years earlier. The UFO in question seemed to be passing over a string of missile sites on that occasion. The Cuban Missile Crisis had occurred only five years previously. "Ban the Bomb" protests were taking place all over the world, and observers in academia and the media regularly were talking about payloads, yields, megatons, and all the arcana of nuclear war. Yet these were trained military personnel—on both sides of the Cold War—who were experiencing serious anomalous phenomena with regard to their missile systems that seemed to be connected to the simultaneous appearances of UFOs over their bases; UFOs that seemed able to override the computer systems in control of the launch sequences.

As Tom's interview subject stated, there were heroes those days on both sides of the Cold War. The United States and the Soviet Union came closer to a shooting war than anyone realized in the 1960s, 1970s, and 1980s. It was only the presence of cooler heads that kept the missiles from being launched, even when it seemed as if something else was interfering with the peace process.

▼   ▼   ▼

The United States is often referred to as a "Christian country," and the statistics would certainly seem to suggest that Christians compose the majority of its population. According to a 2011 Pew Research Center Forum on Religion and Public Life, nearly eighty percent of all Americans are Christian. The Soviet Union, however, was an atheistic country almost by definition. As in the People's Republic of China, membership in a religious organization or community meant that one lost their Party status and were relegated to the "outer darkness" of the dictatorship of the proletariat. When it comes to the Phenomenon, however, it obviously doesn't matter whether one is a religious person or not. The Phenomenon is no respecter of religious affiliation or lack thereof, and when scientific instruments begin reacting in its presence, it's time to do some serious introspection as to what constitutes the "religious" aspect of the Phenomenon and what constitutes its "scientific" aspect, if that duality is even relevant.

But it will be difficult.

▼　　▼　　▼

Attacks on paranormal research sometimes border on the hysterical; facts are ignored or misrepresented by the scientists and the skeptics in order to promote their own philosophy, and these errors are not picked up or corrected by their peers. For some reason, when it comes to the paranormal (and we can include the UFO Phenomenon in that category) it is acceptable for skeptics to be sloppy, to fudge the paperwork, to misrepresent findings, and to ridicule those who sincerely pursue this type of research at risk to their own credibility and careers. The skeptics begin from a conclusion, that the entire field is bunk, and work backward from there, which is the very charge they level at the "believers" whom they accuse of ignoring evidence that is not in line with their claims, whether of paranormal abilities, ancient aliens, or flying saucer crashes. To be sure, both sides are guilty of this type of un-scientific approach, but in the case of the skeptics this failure is more worrisome because it is science itself—and the scientific method—that they are attempting to promote.

Carl Sagan famously insisted that "extraordinary claims require extraordinary proof," which sounds reasonable at first until you unpack it. What is "extraordinary" and what is not? To a scientist this definition may seem self-evident, but there is a hint of dogmatism in this statement that is difficult to identify at first. An extraordinary claim, to Sagan and to the scientists of his generation, is anything— any phenomenon, including especially any *analysis* of that phenomenon—that challenges the accepted wisdom. The implication is that the accepted wisdom has reached the limits of human capability to understand, to experience, to build hypotheses. The laws of physics are immutable laws; as in the example of the church, to suggest transgression of those laws is a sure indication of heresy.

The difference is that the church based its laws on language—on scripture, revelation, and dogma—and not on physical data and certainly not on logic, even though Catholic thinkers like Thomas Aquinas tried to apply logic to theological arguments. Science is based on logic and reason, in which one statement of truth proceeds normally from previous statements, and all are based on observation of physical phenomena and the repeatability of experimentation giving the same results. There is a predictive aspect of science that is lacking in religion, and it is this reliable predictability that gives science its power over other forms of knowledge. Thus, when Sagan talks about extraordinary claims he is referring to observations of phenomena that cannot be associated with previous statements about truth, and particularly about phenomena that cannot be predicted or repeated under laboratory conditions or through, for instance, astronomical observations and calculations. In other words, from a purely scientific perspective, UFOs cannot exist. The propulsion systems are impossible; the physical characteristics of UFOs are impossible; and the idea that life forms use these impossible vehicles to travel impossible distances . . . well, it's all just impossible.

The persistent viewpoint is that Ufologists are not talking about anything "real," a viewpoint that pretty much short-circuits the discussion.

It is for that reason that it becomes necessary to develop an alternate definition of reality, one that includes the Phenomenon, so that we can begin to understand it within a legitimate philosophical framework. It is also necessary to extricate what we know of today's secret weapons and secret technology from what we know of the historical UFOs, as we will do in *Sekret Machines: War*.

We may be surprised to discover a level of interconnection that we never suspected existed.

Bibliography

Adamski, George. *Behind the Flying Saucer Mystery.* New York: Warner Books, 1967, 1974.

Alexander, John B. *UFOs: Myths, Conspiracies, and Realities.* New York: Thomas Dunne Books, 2011.

———. *Reality Denied: Firsthand Experiences with Things that Can't Happen—But Did.* San Antonio, TX: Anomalist Books, 2017.

ASPR Newsletter. "Reception for Ingo Swann." *ASPR Newsletter,* Number 14, September 1972.

Baudrillard, Jean. *Simulacra and Simulation.* Translated by Sheila Faria Glaser. Ann Arbor: University of Michigan Press, 1994.

Bentall, Richard P. "Research into Psychotic Symptoms: Are There Implications for Parapsychologists?" *European Journal of Parapsychology* 15 (2000): pp. 79–88.

Bergson, Henri. *The Two Sources of Morality and Religion.* Westport, CT: Greenwood Press, 1928.

Bower, Tom. *Blind Eye to Murder.* London: Warner Books, 1997.

Breslaw, Elaine. *Tituba: Reluctant Witch of Salem.* New York: New York University Press, 1996.

Bryan, C. D. B. *Close Encounters of the Fourth Kind: A Reporter's Notebook on Alien Abduction, UFOs, and the Conference at M.I.T.* New York: Penguin, 1996.

Bullard, Thomas E. *The Myth and Mystery of UFOs.* Lawrence, KS: University Press of Kansas, 2010.

Chalmers, David. *The Conscious Mind: In Search of a Fundamental Theory.* New York: Oxford University Press, 1996.

Clynes, Manfred E., and Kline, Nathan S. "Cyborgs and Space." *Astronautics,* September 1960: pp. 26–27.

Corkin, Suzanne. "What's New with the Amnesic Patient H.M.?" *Nature Reviews/Neuroscience* 3 (February 2002): pp. 153–159.

Crick, F. H., and Orgel, L. E. "Directed Panspermia." *Icarus* 19 (3): pp. 341–346.

Crick, Francis. *The Astonishing Hypothesis: The Scientific Search for the Soul.* New York: Scribner, 1994.

Crick, Francis, and Koch, Christof. "Towards a Neurobiological Theory of Consciousness." *Seminars in the Neurosciences* 2 (1990): pp. 263–275.

Daniel, Missy. "John E. Mack: The Psychiatrist and Biographer Addresses Human Encounters with Aliens." *Publishers Weekly,* April 18, 1994.

Dawkins, Richard. *The Selfish Gene.* Oxford: Oxford University Press, 1976.

Dehaene, Stanislas. *Consciousness and the Brain.* New York: Penguin, 2014.

Descartes, René. *The Discourse on Method.* Translated by John Veitch. Chicago: Open Court Publishing Company, 1910.

Driscoll, Robert W. "Engineering Man for Space: The Cyborg Study." Final Report, NASw-512, May 15, 1963. Farmingdale, CT: United Aircraft Corporate Systems Center.

Eiseley, Loren. *The Invisible Pyramid*. Lincoln: University of Nebraska Press, 1970, 1998.

———. *The Night Country*. New York: Charles Scribners, 1971.

———. *The Lost Notebooks of Loren Eiseley*. Edited by Kenneth Heuer. Boston: Little, Brown, 1987.

Eliade, Mircea. *The Forge and the Crucible*. Chicago: University of Chicago Press, 1956, 1978.

———. *Shamanism: Archaic Techniques of Ecstasy*. Translated by Willard R. Trask. London: Arkana, 1989.

Emes, Richard D., et al. "Evolutionary Expansion and Anatomical Specialization of Synapse Proteome Complexity." *Nature Neuroscience* 11 (7): pp. 799–806. doi: 10.1038/nn.2135.

Feinberg, Todd E., and Mallatt, John. "The Evolutionary and Genetic Origins of Consciousness in the Cambrian Period over 500 Million Years Ago." *Frontiers in Psychology* 4 (October 2013): Article 667.

Foucault, Michel. *Madness and Civilization*. New York: Vintage, 1965, 1988.

Fuller, John G. *The Interrupted Journey: Two Lost Hours "Aboard a Flying Saucer."* New York: The Dial Press, 1966.

Garrett, Garet. *Ouroboros: Or the Mechanical Extension of Mankind*. London: Kegan Paul, Trench, Trubner & Co., 1926.

Goodfellow, Ian J., et al. "Generative Adversarial Networks." June 10, 2014. https://arxiv.org/abs/1406.2661.

Grandy, John K. "The Three Neurogenetic Phases of Human Consciousness." *Journal of Conscious Evolution*, Issue 9 (2013): p. 2.

Hameroff, Stuart. "A Brief History of (a Study of) Consciousness." Talk given at the Science and Nonduality (SAND) Conference, October 22, 2015.

Hameroff, Stuart, and Penrose, Roger. "Consciousness in the Universe: A Review of the 'Orch OR' Theory." *Physics of Life Reviews* 11 (2014): pp. 39–78.

Hansen, Chadwick. *Witchcraft at Salem.* New York: George Braziller, 1992.

Harris, Sam. "An Atheist Manifesto." December 7, 2005. www.truthdig.com/didg/item/200512_an_atheist_manifesto/.

Hastings, Robert. "UFOs and Nukes." *MUFON Journal,* August 2012.

Heisenberg, Werner. *Physics and Philosophy.* New York: Harper & Row, 1958.

Hind, Angela. "Alien Thinking." *BBC News,* June 8, 2005. http://news.bbc.co.uk/2/hi/uk_news/magazine/4071124.stm.

Hind, Cynthia. "Ariel School Report, Case No. 96." *UFO Afrinews* No. 12 (July 1995): p. 7.

Humphries, Mark. "Your Cortex Contains 17 Billion Computers." February 12, 2018. http://www.systemsneurophysiologylab.manchester.ac.uk. Accessed Feb. 24, 2018.

Jacobs, David M. *Secret Life: Firsthand Accounts of UFO Abductions.* New York: Simon & Schuster, 1992.

Jamaluddin, Ahmad. "Humanoid Encounters in Malaysia." *MUFON Journal* 141 (November 1979).

———. "Mini-Entities at Kuan Air Force Base." *Flying Saucer Review* 26 (5).

———. "Alien Encounters in Southeast Asia." *Alien Encounters* 17 (October 1997).

Jung, Carl G. *Flying Saucers: A Modern Myth of Things Seen in the Sky.* London: Routledge & Kegan Paul, 1959.

Kean, Leslie. *UFOs: Generals, Pilots, and Government Officials Go on the Record.* New York: Harmony Books, 2010.

Kelleher, Colm A. *Hunt for the Skinwalker: Science Confronts the Unexplained at a Remote Ranch in Utah.* New York: Paraview Pocket Books, 2005.

Keyhoe, Donald E. *Aliens from Space: The Real Story of Unidentified Flying Objects.* New York: Signet Books, 1973.

Kline, Nathan S., and Clynes, Manfred. "Drugs, Space, and Cybernetics: Evolution to Cyborgs." In *Psychophysiological Aspects of Space Flight,* edited by Bernard E. Flaherty. New York: Columbia University Press, 1961.

Koestler, Arthur. *The Ghost in the Machine.* New York: Macmillan, 1967.

Koontz, Dean. *One Door Away from Heaven.* New York: Bantam, 2001.

Kripal, Jeffrey J. *Authors of the Impossible: The Paranormal and the Sacred.* Chicago: University of Chicago Press, 2010.

Kuhn, Thomas. *The Structure of Scientific Revolutions.* Chicago: University of Chicago Press, 1962.

Levenda, Peter. *Sinister Forces: Book One.* Walterville, OR: Trine Day, 2005.

———. *Tantric Temples: Eros and Magic in Java.* Lake Worth, FL: Ibis Press, 2011.

Machen, Arthur. "The Great God Pan." In *The Strange World of Arthur Machen.* New York: Juniper Press, 1960.

Mack, John E. *Abduction: Human Encounters with Aliens.* New York: Ballantine Books, 1994.

———. "Messengers from the Unseen." *Oberlin Alumni Magazine,* Winter 2002–03: p. 28.

Mack, John E., and Callimanopulos, Dominique. "The Ariel School Sighting." *CenterPiece Magazine* Spring–Summer 1995: pp. 10–11.

Madrigal, Alexis C. "The Man Who First Said 'Cyborg,' 50 Years Later." *The Atlantic,* September 30, 2010.

Marek, Roger, et al. "Hippocampus-Driven Feed-Forward Inhibition of the Prefrontal Cortex Mediates Relapse of Extinguished Fear." *Nature Neuroscience* 21 (2018): pp. 384–392.

Marks, David, and Kammann, Richard. *The Psychology of the Psychic: A Penetrating Scientific Analysis of Claims of Psychic Abilities.* Buffalo, NY: Prometheus Books, 1980.

Marks, John. *The Search for the "Manchurian Candidate": The CIA and Mind Control.* New York: Times Books, 1979.

Marx, Karl. "Critique of Hegel's Philosophy of Right." *Deutsche-Französische Jahrbücher*, February 1844 (Paris).

Mather, Cotton. *Wonders of the Invisible World.* 1693.

McLuhan, Marshall. *Understanding Media: The Extensions of Man.* New York: McGraw-Hill, 1964.

Meier, C. A., ed. *Atom and Archetype: the Pauli/Jung Letters 1932–1958.* Princeton, NJ: Princeton University Press, 2001.

Minot Daily News. "Minot Launch Control Center 'Saucer' Cited As One Indication of Outer Space Visitors." December 6, 1966.

Morehouse, David. *Psychic Warrior.* New York: St. Martin's Paperbacks, 1996.

Nair, Roshni. "The Truth Is Out There: Tales from India's UFO Investigators." *Hindustan Times*, May 11, 2017.

Neetup. "Alien Abduction Experience in India." http://ireport.cnn.com/docs/DOC-838658. Accessed September 7, 2012.

Nithianantharajah, J., et al. "Synaptic Scaffold Evolution Generated Components of Vertebrate Cognitive Complexity." *Nature Neuroscience* 16 (1): 16–24. doi: 10.1038/nn.3276.

Oizumi, Masafumi, et al. "From the Phenomenology to the Mechanisms of Consciousness: Integrated Information Theory 3.0." *PLOS Computational Biology* 10, No. 5 (May 2014).

Onfray, Michel. *The Atheist Manifesto: The Case Against Christianity, Judaism, and Islam.* Melbourne, Australia: Melbourne University Press, 2007.

Oppenheimer, J. Robert. *Science and the Common Understanding.* New York: Simon & Schuster, 1954.

Paez, Danny. "SXSW: Ray Kurzweil Says DNA Is 'Outdated Software' Biotech Can 'Reprogram.'" Interview at SXSW 2018, March 13, 2018.

Peat, F. David. *Synchronicity: The Bridge Between Matter and Mind.* New York: Bantam Books, 1987.

Peek, Philip M. *African Divination Systems: Ways of Knowing.* Bloomington: Indiana University Press, 1991.

Pribram, Karl H. *Brain and Perception: Holonomy and Structure in Figural Processing.* Hillsdale, NJ: Lawrence Erlbaum Assoc., 1991.

Puharich, Andrija. *Beyond Telepathy.* New York: Anchor Press/ Doubleday, 1973.

———. *URI: A Journal of the Mystery of Uri Geller.* New York: Anchor Press, 1974.

Puthoff, H. E. "CIA-Initiated Remote Viewing at Stanford Research Institute." *Intelligencer: A Journal of U.S. Intelligence Studies,* Summer 2001: pp. 60–67.

Puthoff, H. E., et al. "Inflation-Theory Implications for Extraterrestrial Visitation." JBIS 58 (2005): pp. 43–50.

Reuters. "Huge UFOs Sighted in Malaysia." October 18, 1995.

Sagan, Carl. *The Demon-Haunted World: Science as a Candle in the Dark.* London: Headline Books, 1997.

Schnabel, Jim. *Remote Viewers: The Secret History of America's Psychic Spies.* New York: Dell, 1997.

shCherbak, Vladimir I., and Makukov, Maxim A. "The 'WOW Signal' of the Terrestrial Genetic Code." *Icarus* 224 (1): pp. 228–242.

Snefjella, Bryor. "This Is the A-Bomb Moment for Computer Science." *Buzzfeed,* March 22, 2018.

Stanislavski, Konstantin. *An Actor Prepares.* New York: Routledge/ Theatre Arts Books, 1936.

Strieber, Whitley. *Communion: A True Story.* New York: Avon Books, 1987.

Strieber, Whitley, and Kripal, Jeffrey J. *The Super Natural: A New Vision of the Unexplained.* New York: Penguin Random House, 2016.

Strieber, Whitley, and Strieber, Anne. *The Afterlife Revolution.* San Antonio, TX: Walker & Collier, 2017.

Summers, Montague. *The Malleus Maleficarum of Heinrich Kramer and James Sprenger.* New York: Dover, 1971.

Swann, Ingo. *Penetration: The Question of Extraterrestrial and Human Telepathy.* Rapid City, SD: Ingo Swann Books, 1998.

Swords, Michael, and Powell, Robert. *UFOs and Government: A Historical Inquiry.* San Antonio, TX: Anomalist Books, 2012.

Temple, Robert. "The Prehistory of Panspermia: Astrophysical or Metaphysical?" *The International Journal of Astrobiology* 6 (2): pp. 169–180.

Tononi, Giulio, et al. "Integrated Information Theory: From Consciousness to Its Physical Substrate." *Nature Reviews: Neuroscience* 17 (July 2016): pp. 450–461.

Trachtenburg, Joshua. *Jewish Magic and Superstition: A Study in Folk Religion.* New York: Atheneum, 1975.

Turner, Patricia A. *I Heard It Through the Grapevine: Rumor in African-American Culture.* Berkeley: University of California Press, 1993.

Vallée, Jacques. *Forbidden Science: Journals 1957–1969.* Berkeley, CA: North Atlantic Books, 1992.

———. *Forbidden Science 2: California Hermetica.* San Antonio, TX: Anomalist Books, 2008, 2017.

———. *Forbidden Science 3: On the Trail of Hidden Truths.* San Antonio, TX: Anomalist Books, 2012.

Vandervert, Larry R. "Chaos Theory and the Evolution of Consciousness and Mind: A Thermodynamic–Holographic Resolution to the Mind-Body Problem." *New Ideas in Psychology* 13, No. 2 (1995): pp. 107–127.

Government Documents

These are representative examples of the vast library of declassified CIA, DIA, and US military files (as well as Russian, Chinese, and Japanese files held in various archives) on Psi research, remote viewing programs, and SRI and related subjects that we used as background for this work. These files are available for free download from the respective government websites. Often, these documents either do not have an "author" or the name of the creator of the document has been redacted. We have listed a few of these documents with their CIA locator number as well as a brief description. Interested researchers are encouraged to download the complete sets of files from the appropriate websites, as they are educational and revealing.

▼　　▼　　▼

Anon., "Memorandum for the Record; Subject: Briefing for House Permanent Select Committee Re: OTS Involvement in ESP Research," OTS/BAB Memo #287-77, November 17, 1977. This seven-page memo was originally classified as SECRET and provides a summary of attempts by OTS (Office of Technical Services, a department within CIA) to develop a viable intelligence capability using remote viewing in collaboration with SRI and in particular with Pat Price, although Ingo Swann is also mentioned. It refers to the potentially "embarrassing" aspect of the research, particularly at a time when CIA and OTS were subjects of investigation concerning the Watergate affair. CIA-RDP96-00787R000500170001-0

Approved for Release 2005/03/24.

Anon., "Project GRILL FLAME," August 20, 1981. This is an eight-page DIA memo on the remote viewing project known as GRILL FLAME and consists of a question and an answer on each page, accompanied by handwritten notes. It references CIA, SRI, and INSCOM, and discusses financial budgeting issues as well. One handwritten notation reads: "To me the spirit of the question indicates what are you doing to assess the Phenomenon?" Another reads: "Army guidance to O'Keefe: Tell them as little as possible." (!) CIA-RDP96-00788R001100210002-6

Approved for Release 2005/03/24.

Anon., "Project STAR GATE," undated. This thirty-six page document was classified Secret NoForn Star Gate Limdis (Secret, No Foreigners, Star Gate, Limited Distribution), indicating its nature as a Special Access Program. It provides useful definitions of topics such as ESP, Psychokinesis, Remote Viewing, etc. CIA-RDP96-00789R003300210001-2

Approved for Release 2000/08/08.

Puthoff, Harold E., and Targ, Russell, "Standard Remote-Viewing Protocol: Local Targets," November 1978 (revised). SRI International. This is a sixteen-page document that goes over the protocols for remote viewing testing procedures, written by the very people who created the remote viewing program at SRI, which eventually morphed into the various Projects— GRILL FLAME, STAR GATE, etc.—that were funded by the CIA, DIA, and Army over the course of some decades. CIA-RDP96-00788R001300050001-3

Approved for Release 2000/08/07.

Endnotes

INTRODUCTION TO SECTION ONE

1   Loren Eiseley, Kenneth Heuer, ed., *The Lost Notebooks of Loren Eiseley*, Boston, Little Brown, 1987, p. 199.

2   Loren Eiseley, *The Invisible Pyramid*, Lincoln, University of Nebraska Press, 1970, pp. 54–55.

CHAPTER ONE

3   Angela Hind, "Alien Thinking," BBC News, 2005/06/08, http://news.bbc.co.uk/go/pr/fr/-/2/hi/uk_news/magazine/4071124.stm. Retrieved January 2016.

4   Dr. J. Allen Hynek, Speech before the United Nations, November 27, 1978.

5   Dr. J. Allen Hynek, Speech before the United Nations, November 27, 1978.

6   John G. Fuller, *The Interrupted Journey: Two Lost Hours "Aboard a Flying Saucer,"* The Dial Press, New York, 1966, from which the basic outline of this story is derived.

7   John Mack quoted in Missy Daniel, "John E. Mack: The Psychiatrist and Biographer Addresses Human Encounters with Aliens," *Publishers Weekly*, 18 April 1994, pp. 40–41.

8   Dr. James McDonald was a physicist who testified before the US Congress in 1968 on the reality of the UFO Phenomenon and was ridiculed mercilessly for it. He took his own life in 1971.

9   As an aside, one may note that a biomarker such as a burn from too-close contact with the Phenomenon may have been referenced in the Bible, in Exodus 34, where Moses comes down from Mount Sinai after spending time with God, unaware that his face is shining so brightly that he is forced to wear a veil to cover it. The Hebrew word for this condition—קָרַן—is so rare that it appears nowhere else in the Bible.

10   As another aside, one of the most famous individuals associated with the alleged New Orleans–based conspiracy to assassinate President Kennedy—David Ferrie—also suffered from alopecia, which is why he often pasted on fake eyebrows as well as wore a wig. Ferrie had been a pilot for Eastern Airlines, but we do not know why or how he developed alopecia. This is only of interest because two other characters from the conspiracy cabal—former FBI SAC Guy Banister and Fred Crisman—had a background in UFOs dating from 1947 and the Kenneth Arnold and Maury Island affairs. See Levenda, *Sinister Forces: The Nine*, for more information on this bizarre series of coincidences.

11   Dr. Hal Puthoff, presentation before the SSE/IRVA (Society for Scientific Exploration/International Remote Viewing Association), June 8, 2018, in Las Vegas, NV. Transcript available on the website of the Paradigm Research Institute, http://paradigmresearchgroup.org/wordpress/2018/06/12/dr-hal-puthoff-presentation-at-the-sse-irva-conference-las-vegas-nv-15-june-2018/, Retrieved June 30, 2018.

CHAPTER TWO

12   This was the famous Byelokoroviche incident at an IRBM (Intermediate Range Ballistic Missile) base on October 4, 1982.

13   This was the equally famous Minot AFB incident of the late 1960s. See the Robert Hastings article on "UFOs and Nukes" in the *MUFON Journal*, August 2012, or his book by the same name, for details on these events.

14   For instance, in the Discovery Channel program *Into the Universe with Stephen Hawking*, April 2010.

15   Although there are those in Congress and the military establishment who are not so sure!

16   See for instance Elaine Breslaw, *Tituba: Reluctant Witch of Salem*, New York University Press, New York, 1996 and Chadwick Hansen, *Witchcraft at Salem*, George Braziller, New York, 1992.

17   Montague Summers, "Introduction to 1948 Edition," *The Malleus Maleficarum of Heinrich Kramer and James Sprenger*, Dover, New York, 1971, p. v.

CHAPTER THREE

18   Joshua Trachtenberg, *Jewish Magic and Superstition: A Study in Folk Religion*, Atheneum, New York, 1975, p. 85.

19   Genesis Rabbah, 65:21.

20   St. John Chrysostom, *Homilies*, 1 Corinthians 32:6: "He does not say these things as if he attributed to angels knees and bones."

21   C. D. B. Bryan, *Close Encounters of the Fourth Kind: A Reporter's Notebook on Alien Abduction, UFOs, and the Conference at M.I.T.*, Penguin, New York, 1996, p. 28.

22   C. D. B. Bryan (1996), p. 28.

23   C. D. B. Bryan (1996), pp. 28–29.

24   See for instance the citation from Berossus in Cory, Isaac Preston, *TThe Ancient Fragments; Containing What Remains of the Writings of Sanchoniatho, Berossus, Abydenus, Megasthenes, and Manetho* (trans. by E. Richmond Hodges), 1876, London, Reeves & Turner, p. 57.

25   See for instance Joshua Trachtenberg, *Jewish Magic and Superstition: A Study in Folk Religion*, Atheneum, New York, 1975, pp. 84–86.

26   *The Sodei Raza of Eleazar of Worms* (1165–1230) is one of the early examples.

27   *Tractate Sanhedrin*, 38b.

28   John Marks, *The Search for the "Manchurian Candidate": The CIA and Mind Control*, Times Books, New York, 1979, p. 211.

29   Jacques Vallée, *Forbidden Science: Journals 1957–1969*, North Atlantic Books, Berkeley, 1992, p. 422.

30   See John Marks (1979) as well as Tom Bower, *Blind Eye to Murder*, Warner Books, London, 1997, and Peter Levenda, *Sinister Forces: Book One*, Trine Day, Walterville, OR, 2005.

31   Ingo Swann, *Penetration: The Question of Extraterrestrial and Human Telepathy*, Ingo Swann Books, Rapid City, SD, 1998.

32   Garry P. Nolan, et al., "Whole-genome sequencing of Atacama skeleton shows novel mutations linked with dysplasia," in *Genome Research*, Cold Spring Harbor Laboratory Press, April 2018.

33   Danny Paez, "SXSW: Ray Kurzweil says DNA is 'Outdated Software' Biotech can 'Reprogram,'" March 13, 2018.

CHAPTER FOUR

34   Mircea Eliade, *The Forge and the Crucible*, University of Chicago Press, Chicago, 1956, 1978.

35   Genesis 3:16, "I will greatly increase your pains in childbearing; with pain you will give birth to children."

36   See, for instance, Jasper Hamill, "Pentagon and MoD officials feared UFOs were either 'demonic' or sent by God, former investigators reveal," 8 May 2018, http://metro.co.uk/2018/05/08/religious-pentago-mod-officials-thought-ufos-demonic-divine-former-government-investigators-reveal-7529174/. Retrieved June 1, 2018.

37 As an aside, it is impossible to ignore the fact that science in the twentieth century (and beyond) was dramatically affected by World War II to the extent that our scientists and their most brilliant discoveries came about as a result of wartime experience and work within a wartime environment.

38 Robert Temple, "The prehistory of panspermia: astrophysical or metaphysical?," *The International Journal of Astrobiology*, 6 (2): pp. 169–180 (2007).

39 Peter Levenda, *Tantric Temples: Eros and Magic in Java*, Ibis Press, Lake Worth, FL, 2011, pp. 70–71.

40 Crick, F. H., and Orgel, L. E. (1973), "Directed Panspermia," *Icarus*, 19 (3): pp. 341–346.

41 Vladimir I. *sh*Cherbak and Maxim A. Makukov, "The 'WOW signal' of the terrestrial genetic code," *Icarus*, 6 March 2013, pp. 228–242.

42 Ibid., p. 232.

CHAPTER FIVE

43 J. Robert Oppenheimer, *Science and the Common Understanding*, Simon & Schuster, NY, 1954.

44 This system is also known as sikidy in Madagascar, and by other names in many other African countries. See for instance Philip M. Peek, ed., *African Divination Systems: Ways of Knowing*, Indiana University Press, Bloomington, 1991, pp. 52–68.

45 Werner Heisenberg, *Physics and Philosophy*, New York, Harper & Row, 1958, p. 81.

INTRODUCTION TO SECTION TWO

46 Dean Koontz, *One Door Away from Heaven*, Bantam, NY, 2001, p. 472.

CHAPTER SIX

47   Francis Crick and Christof Koch, "Towards a neurobiological theory of consciousness," *Seminars in the Neurosciences*, Volume 2, 1990: pp. 263–275.

48   Ibid., p. 263.

49   Ibid., p. 264.

50   Ibid., p. 264.

51   René Descartes, *The Discourse on Method*, Part 6.

52   Crick and Koch, p. 264.

53   Crick and Koch, p. 264.

54   That said, for those with an eidetic (or "photographic") memory it may be possible to "read" all the titles at once, but they may not be accessible at once except through memory, which is a different aspect of consciousness and one that we will get to presently.

55   Crick and Koch, p. 270.

56   Suzanne Corkin, "What's new with the amnesic patient H.M.?," *Nature Reviews/Neuroscience*, Volume 3, February 2002, pp. 153–159.

57   Crick and Koch, p. 269.

58   Ibid.

59   Crick and Koch, p. 269. Emphasis in original.

60   Francis Crick, *The Astonishing Hypothesis: The Scientific Search for the Soul*, Scribner, NY, 1994.

61   John K. Grandy, "The Three Neurogenetic Phases of Human Consciousness," *Journal of Conscious Evolution*, Issue 9: 2013, p. 2.

62   Ibid. (Grandy's emphasis.)

63   Ibid.

64   Ibid.

CHAPTER SEVEN

65   Henri Bergson, *The Two Sources of Morality and Religion*, Greenwood Press, Westport, CT, 1928, p. 268.

CHAPTER EIGHT

66   Original report by Lionel Feuillet et al., *The Lancet*, Volume 370, p. 262; referenced in *The New Scientist*, 20 July 2007.

67   Mark Humphries, "Your Cortex Contains 17 Billion Computers," http://www.systemsneurophysiologylab.manchester.ac.uk, Feb. 12, 2018. Retrieved Feb. 24, 2018.

CHAPTER NINE

68   Todd E. Feinberg and John Mallatt, "The evolutionary and genetic origins of consciousness in the Cambrian Period over 500 million years ago," *Frontiers in Psychology*, October 2013, Volume 4, Article 667.

69   Francis Crick and Christof Koch, "Towards a neurobiological theory of consciousness," *Seminars in Neuroscience*, Volume 2, 1990: pp. 263–275.

70   J. Nithianantharajah et al., "Synaptic scaffold evolution generated components of vertebrate cognitive complexity," *Nature Neuroscience* 2013 January, 16(1): 16–24, doi:10.1038/nn.3276; and Richard D. Emes et al., "Evolutionary expansion and anatomical specialization of synapse proteome complexity," *Nature Neuroscience* 2008 July, 11(7): 799–806. doi:10.1038/nn.2135.

71   There is no consensus today on whether Pikaia was a vertebrate or invertebrate.

72   MAGUK is the acronym for "membrane-associated guanylate kinases," a "superfamily" of proteins.

73   Except, of course, in cases of Near Death Experience or NDE, in which a "dead" person comes back to life in the operating room after having been dead for a short time, with all their memories and self-awareness intact. This is called a "Near Death Experience" because of the basic medical assumption that since the person came back to life they were never really dead in the first place, only "near" death. The logic of this assumption is, of course, debatable.

74   "Quantum mechanics explains efficiency of photosynthesis," January 9, 2014, https://phys.org/news/2014-01-quantum-mechanics-efficiency-photosynthesis.html.

75   A nanometer is one billionth of a meter.

76   Stuart Hameroff and Roger Penrose, "Consciousness in the universe: A review of the 'Orch OR' theory," *Physics of Life Reviews*, 11 (2014), pp. 39–78.

77   Ibid., p. 73.

78   Ibid., pp. 39–78.

79   Stuart Hameroff, "A brief history of (a study of) consciousness," a talk given on October 22, 2015, at the Science and Nonduality (SAND) Conference.

CHAPTER TEN

80   Larry R. Vandervert, "Chaos Theory and the Evolution of Consciousness and Mind: A Thermodynamic–Holographic Resolution to the Mind-Body Problem," *New Ideas in Psychology*, 1995, Vol. 13, No. 2, pp. 107–127.

81   See for instance Karl H. Pribram, *Brain and Perception: Holonomy and Structure in Figural Processing*, Mahwah, NJ, Lawrence Erlbaum Assoc., 1991.

82   Vandervert (1995), p. 114.

83   Ibid. p. 116.

84   Ibid., p. 116.

85   Ibid, p. 113.

86   Ibid, p. 113.

87   Ibid, p. 115.

88   Stanislas Dehaene, *Consciousness and the Brain*, Penguin, New York, 2014, p. 263.

89   Ibid., p. 88.

90   Ibid., p. 153.

91   Ibid., p. 153.

INTRODUCTION TO SECTION THREE

92   Konstantin Stanislavski, *An Actor Prepares*, Routledge/Theatre Arts Books, New York, 1936, 1989, p. 176.

93   Michel Foucault, *Madness and Civilization*, Vintage, NY, 1965, 1988, p. 100.

CHAPTER ELEVEN

94   David Chalmers, *The Conscious Mind: In Search of a Fundamental Theory*, Oxford University Press, New York, 1996.

95   Ian J. Goodfellow, et al., "Generative Adversarial Nets," arXiv:1406.2661 [stat.ML], 10 June 2014.

96   Bryor Snefjella, "This is the A-Bomb Moment for Computer Science," *Buzzfeed*, March 22, 2018.

97   See, for instance, Giulio Tononi, Melanie Boly, Marcello Massimini, and Christof Koch, "Integrated information theory: from consciousness to its physical substrate," *Nature Reviews: Neuroscience*, July 2016, Volume 17, pp. 450–461; and Masafumi Oizumi, Larissa Albantakis, and Giulio Tononi, "From the phenomenology to the mechanisms of consciousness: Integrated information theory 3.0," *PLOS Computational Biology*, May 2014, Volume 10, Issue 5.

98 See, for instance, the gun camera footage of a UFO spotted off the coast of southern California in November 2004 that was released in December 2017 in a much-remarked *New York Times* article, which broke the news that the US government had been studying UFOs for years. Helene Cooper, Ralph Blumenthal, and Leslie Kean, "Glowing Auras and 'Black Money': The Pentagon's Mysterious U.F.O. Program," *New York Times*, December 16, 2017.

CHAPTER TWELVE

99 Manfred E. Clynes and Nathan S. Kline, "Cyborgs and Space," *Astronautics*, September 1960, pp. 26–27.

100 We will look at this event more closely in *Sekret Machines: War.*

101 Clynes and Kline, "Cyborgs and space." Emphasis in original.

102 Ibid.

103 Ibid.

104 Norbert Wiener, *Cybernetics, or Control and Communication in the Animal and the Machine*, Cambridge, MA, MIT Press, 1948.

105 Driscoll, Robert W., "Engineering Man for Space: The Cyborg Study," Final Report, NASw-512, May 15, 1963, United Aircraft Corporate Systems Center, Farmingdale, CT.

106 Nathan S. Kline and Manfred Clynes, "Drugs, Space, and Cybernetics: Evolution to Cyborgs," *Psychophysiological Aspects of Space Flight*, Columbia University Press, New York, 1961. Emphasis in original.

107 Ibid.

108 Ibid.

109 Ibid.

110 Ibid.

111 Ibid.

112   Alexis C. Madrigal, "The Man Who First Said 'Cyborg,' 50 Years Later," *The Atlantic*, Sept. 30, 2010.

CHAPTER THIRTEEN

113   Ingo Swann, *Penetration: The Question of Extraterrestrial and Human Telepathy*, Ingo Swann Books, Rapid City, SD, 1998, p. 175. Emphasis in original.

114   Jacques Vallée, *Forbidden Science 2: California Hermetica*, Anomalist Books, San Antonio, TX, 2008, 2017, p. 409.

115   In fact, as this book was being completed, retired Army Colonel John B. Alexander, PhD—who has written on the UFO Phenomenon and on Skinwalker Ranch as well as on other paranormal subjects—gave a talk at the Rhine Center in North Carolina on March 24, 2018, about his career and his work concerning the paranormal.

116   Zener cards consisted of a set of five symbols printed on cards, one symbol to a card in a shuffled deck consisting of perhaps 20 such cards, and test subjects had to determine which symbol the experimenter was looking at, one by one, until all cards were used. The percentage of correct guesses was an indicator of Psi if it exceeded probability.

117   H. E. Puthoff, "CIA-Initiated Remote Viewing at Stanford Research Institute," *Intelligencer: Journal of U.S. Intelligence Studies*, Summer 2001, pp. 60–67.

118   INSCOM, or US Army Intelligence and Security Command, was founded in 1977.

119   Note dated 11/2/78 from Hal Puthoff to [redacted]: "Just a quick note. For what it's worth, Swann now has a TS Clearance with DOD." Approved for Release 2003/09/10: CIA-RDP96-00787R000200020029-9.

120   "Reception for Ingo Swann," *ASPR Newsletter*, Number 14, September 1972, New York.

121   Memorandum for the Record, Subject: Final Report of Parapsychology Studies, Stanford Research Institute, 30 August 1973. [Author redacted.] Approved for Release 2003/09/10: CIA-RDP96-00787R000200010008-3.

122   "Negentropy" refers to reverse entropy, or becoming order. Entropy works in the opposite direction—disorder. "EPR" refers to the Einstein-Podolsky-Rosen Paradox, a critique of quantum mechanics.

123   C. A. Meier, ed., *Atom and Archetype: The Pauli/Jung Letters 1932–1958*, Princeton University Press, Princeton (NJ), 2001. The subject of UFOs comes up in this collection of correspondence, with Jung wondering if the Phenomenon is purely psychological, since it seems to mirror the insecurities of the population.

124   See, for instance, F. David Peat, *Synchronicity: The Bridge Between Matter and Mind*, Bantam Books, New York, 1987.

125   [Author redacted], "Quantum Physics and Parapsychology: An interpretative conference report containing some comments about prospects in parapsychological research," October 28, 1974, Approved for release 2003/09/09: CIA-RDP96-00787R000200180001-2. Emphasis in original.

126   "Quantum Physics and Parapsychology," 1974.

127   Ibid.

128   Ibid.

129   "Project Star Gate," Defense Intelligence Agency, (undated) Secret, Noforn, Star Gate, Limdis, Approved for Release 2000/08/08: CIA-RDP96-00789R003300210001-2.

130   "Center Lane Personnel Selection Procedures (U)" dated 10 October 1984, and included in Center Lane Information Papers for Director, Defense Intelligence Agency, 7 March 1985. Approved for Release 2007/12/20: CIA-RDP96-00788R000400060001-2.

131   Jim Schnabel, *Remote Viewers: The Secret History of America's Psychic Spies*, Dell, NY, 1997, pp. 158–159.

132   Mircea Eliade, *Shamanism: Archaic Techniques of Ecstasy*, Arkana, Penguin Books, 1989, p. 27.

133   Richard P. Bentall, "Research into Psychotic Symptoms: Are There Implications for Parapsychologists?," *European Journal of Parapsychology*, 2000, pp. 15, 79–88.

134   Ibid.

135   Ibid.

136   John E. Mack, "Messengers from the Unseen," *Oberlin Alumni Magazine*, Winter 2002-03, p. 28.

137   See for instance Patricia A. Turner, *I Heard It Through the Grapevine: Rumor in African-American Culture*, University of California Press, Berkeley, 1993, pp. 67–70.

138   Ibid.

139   In April 2018 he was awarded the Dinsdale Award of the Society for Scientific Exploration, "For the application of sound scientific principles and methodologies to the study of remote perception, quantum zero-point fluctuations, and unidentified aerial objects, and for realizing the potential usefulness of these often-shunned phenomena in the real world."

140   Carl Gustav Jung, *Flying Saucers: A Modern Myth of Things Seen in the Sky*, Routledge & Kegan Paul, London, 1959.

141   C. A. Meier, "Letter from Max Knoll to Pauli Concerning UFOs," dated September 12, 1957, Munich, in *Atom and Archetype: The Pauli/Jung Letters 1932–1958*, Princeton University Press, Princeton, 2001, p. 201.

142   Arthur Koestler, *The Ghost in the Machine*, Macmillan, New York, 1967.

143   Marshall McLuhan, *Understanding Media: The Extensions of Man*, McGraw-Hill, NY, 1964.

144   Ingo Swann, *Penetration*, 1998, p. 178.

145   http://www.kurzweilai.net/intelligence-augmentation-device-lets-users-speak-silently-with-a-computer-by-just-thinking. Retrieved April 6, 2018.

146   Roger Marek et al., "Hippocampus-driven feed-forward inhibition of the prefrontal cortex mediates relapse of extinguished fear," *Nature Neuroscience*, Volume 21, pp. 384–392 (2018).

147   Jacques Vallée, *Forbidden Science 3: On the Trail of Hidden Truths*, Anomalist Books, San Antonio, 2012, p. 65.

CHAPTER FOURTEEN

148   Loren Eiseley, *The Invisible Pyramid*, University of Nebraska Press, Lincoln, 1970, 1998, pp. 54–55.

149   Neetup, "Alien Abduction Experience in India," http://ireport.cnn.com/docs/DOC-838658, Retrieved Sept. 7, 2012.

150   Roshni Nair, "The truth is out there: Tales from India's UFO investigators," *Hindustan Times*, May 11, 2017.

151   Ibid.

152   Ahmad Jamaludin, "Mini-entities at Kuan Air Force Base," *Flying Saucer Review*, Vol. 26, No. 5, January 1981 (UK).

153   Ahmad Jamaludin, "Humanoid encounters in Malaysia," *MUFON Journal*, No. 141, November 1979.

154   Reuters, "Huge UFOs Sighted in Malaysia," October 18, 1995.

155   Ahmad Jamaludin, "Alien Encounters in Southeast Asia," *Alien Encounters*, Number 17, October 1997.

156   John E. Mack and Dominique Callimanopulos, "The Ariel School sighting," *CenterPiece Magazine*, Spring-Summer 1995, pp. 10–11.

157   Cynthia Hind, "Ariel School Report, Case No. 96," *UFO Afrinews*, No. 12, July 1995, Harare (Zimbabwe), p. 7.

APPENDIX

158 Michel Onfray, *The Atheist Manifesto: The Case Against Christianity, Judaism, and Islam*, Melbourne University Press, Melbourne, 2007, p. 81.

159 Karl Marx, "Critique of Hegel's Philosophy of Right," *Deutsche-Französische Jahrbücher*, Paris, February 1844.

160 Sam Harris, "An Atheist Manifesto," www.truthdig.com/dig/item/200512_an_atheist_manifesto/, December 7, 2005. Retrieved June 5, 2015.

161 Sam Harris (2005).

162 Isaiah 40:4

163 Carl Sagan, *The Demon-Haunted World: Science as a Candle in the Dark*, Headline Books, London, 1997, p. 70.

164 This information comes primarily from two sources: the report by Robert Hastings, author of *UFOs and Nukes: Extraordinary Encounters at Nuclear Weapons Sites*, who was in direct communication with Schuur in 2007 and who broke this story in an article for the *MUFON Journal* (August 2012), and Tim Hebert's website, http://timhebert.blogspot.com/2013/06/the-passing-of-david-schuur.html. Retrieved June 6, 2015.

165 "Minot Launch Control Center 'Saucer' Cited As One Indication of Outer Space Visitors," *Minot Daily News*, December 6, 1966.

Index

fanatical, 392

objections to study of the Phenomenon, 381

paranormal association with religion, 397

political agenda, 396

religion and, 384–385, 395, 397

science and religion and, 390, 391–394

Soviet Union and, 407

attention, 180–181

auditory sense, 173, 174

autonomic nervous system (ANS), 200, 201

awareness

of AD patients, 188

conscious, neurobiological systems below, 167

self-awareness, 172, 188

unconscious, of DNA structure, 159

awareness and attention, 177, 179, 180, 181, 187, 275, 303

axons, 209, 210

B

Babylonian tradition, 75, 90, 112

Bergson, Henri, 191

binding, 175–176, 177

"Biological SETI," 114, 115

biological symbiosis, 302–303

biomarkers, 32–33

"black box" of consciousness, 166, 275

Blavatsky, Helena, 59, 317

Bluebird program, 320

body, the

in alien abduction, 141

binary energy systems within, 129

in the brain, 256

brain creation of experience of, 255–256

consciousness and, 220–221

death of, 149

left and right sides of, 202

mind and, 149

Bohm, David, 331

brain. *See also* neurons

amygdala, 203–204, 347

cerebellum, 204

cerebrum, 201–202

as classical computer, 256–257

consciousness and, 195, 206, 207–208, 211, 215, 241

environmental impact on, 211

hypothalamus, 203

limb and senses development and, 221–222

medulla, 205

neural pathways, 173, 210–211, 377

as quantum computer, 257

as receiver of consciousness, 195

restructuring in response to trauma, 347

"second," 201

thalamus, 202–203

brain-computer interface, xiv, xv, 271

brain genes, 219

brain stem, 204–205

breathing, 302, 303

K

L

M

ABOUT THE AUTHORS

Tom DeLonge is an award-winning American musician, producer, and director best known as the lead vocalist, guitarist, and songwriter for the platinum-selling rock bands Blink-182 and Angels & Airwaves. His home is San Diego, California, where he focuses on creating entertainment properties that cross music, books, and film with his company To The Stars... Check out his other multi-media projects at ToTheStars.Media.

▼   ▼   ▼

Peter Levenda has an MA in Religious Studies and Asian Studies from FIU, and speaks a variety of languages (some of them dead). He has appeared in numerous television programs for the History Channel, the Discovery Channel, National Geographic, and TNT. He has interviewed Nazis, Klansmen, occultists, CIA officers, and Islamic terrorists in the course of his research, and has visited Chinese prisons and military bases; the Palestine Liberation Organization; the former KGB headquarters (Dzherzinsky Square) in Moscow; and once was celebrated with a state dinner in Beijing.

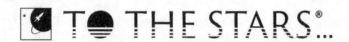

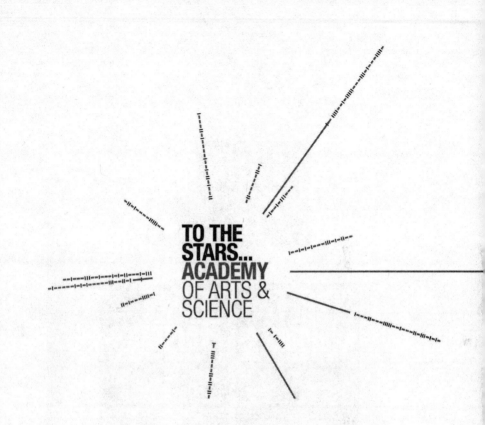

TO THE
STARS...
ACADEMY
OF ARTS &
SCIENCE

A REVOLUTIONARY COLLABORATION BETWEEN ACADEMIA, INDUSTRY,
AND POP-CULTURE FOR SCIENTIFIC ADVANCEMENT.

VISIT **WWW.TOTHESTARSACADEMY.COM**